MECHANICS AND PHYSICS OF ENERGY DENSITY

ENGINEERING APPLICATION OF FRACTURE MECHANICS

Editor-in-Chief: George C. Sih

1. G.C. Sih and L. Faria (eds.), Fracture mechanics methodology: Evaluation of structural components integrity. 1984.
    ISBN 90-247-2941-6.
2. E.E. Gdoutos, Problems of mixed mode crack propagation. 1984.
    ISBN 90-247-3055-4.
3. A. Carpinteri and A.R. Ingraffea (eds.), Fracture mechanisms of concrete: Material characterization and testing. 1984.
    ISBN 90-247-2959-9.
4. G.C. Sih and A. DiTommaso (eds.), Fracture mechanics of concrete: Structural application and numerical calculation. 1984.
    ISBN 90-247-2960-2.
5. A. Carpinteri, Mechanical damage and crack growth in concrete: Plastic collapse to brittle fracture. 1986.
    ISBN 90-247-3233-6.
6. J.W. Provan (ed.), Probabilistic fracture mechanics and reliability. 1987.
    ISBN 90-247-3334-0.
7. A.A. Baker and R. Jones (eds.), Bonded repair of aircraft structures. 1987.
    ISBN 90-247-3606-4.
8. J.T. Pindera and M.-J. Pindera, Isodyne stress analysis. 1989.
    ISBN 0-7923-0269-9.
9. G.C. Sih and E. E. Gdoutos (eds.), Mechanics and physics of energy density. 1992.
    ISBN 0-7923-0604-X.
10. E.E. Gdoutos, Fracture mechanics criteria and applications. 1990.
    ISBN 0-7923-0605-8.
11. G.C. Sih, Mechanics of fracture initiation and propagation. 1991.
    ISBN 0-7923-0877-8.

Mechanics and Physics of Energy Density

Characterization of material/structure behavior with and without damage

George C. Sih

Institute of Fracture and Solid Mechanics
Lehigh University
Bethlehem, Pennsylvania, USA

and

Emmanuel E. Gdoutos

Department of Civil Engineering
Democritus University of Thrace
Xanthi, Greece

Kluwer Academic Publishers

Dordrecht / Boston / London

Library of Congress Cataloging-in-Publication Data

Mechanics and physics of energy density : characterization of
  material/structure behavior with and without damage / [edited by]
  George C. Sih and Emmanuel E. Gdoutos.
       p.   cm. -- (Engineering application of fracture mechanics ; 9)
    Proceedings of a symposium held July 17-20, 1989, at the
Democritus University of Thrace.
    Includes bibliographical references
    ISBN 0-7923-0604-X
    1. Fracture mechanics--Congresses. 2. Strains and stresses-
-Congresses.   I. Sih, G. C. (George C.) II. Gdoutos, E. E., 1948-
. III. Series.
TA409.M399 1990
620.1'126--dc20   90-41854

ISBN 0-7923-0604-X

_____

Published by Kluwer Academic Publishers,
P.O. Box 17, 3300 AA Dordrecht, The Netherlands.

Kluwer Academic Publishers incorporates the publishing programmes of
D. Reidel, Martinus Nijhoff, Dr W. Junk and MTP Press.

Sold and distributed in the U.S.A. and Canada
by Kluwer Academic Publishers,
101 Philip Drive, Norwell, MA 02061, U.S.A.

In all other countries, sold and distributed
by Kluwer Academic Publishers Group,
P.O. Box 322, 3300 AH Dordrecht, The Netherlands.

printed on acid-free paper

Contents

Series on engineering application of fracture mechanics

Fracture mechanics technology has received considerable attention in recent years and has advanced to the stage where it can be employed in engineering design to prevent against the brittle fracture of high-strength materials and highly constrained structures. While research continued in an attempt to extend the basic concept to the lower strength and higher toughness materials, the technology advanced rapidly to establish material specifications, design rules, quality control and inspection standards, code requirements, and regulations of safe operation. Among these are the fracture toughness testing procedures of the American Society of Testing Materials (ASTM), the American Society of Mechanical Engineers (ASME) Boiler and Pressure Vessel Codes for the design of nuclear reactor components, etc. Step-by-step fracture detection and prevention procedures are also being developed by the industry, government and university to guide and regulate the design of engineering products. This involves the interaction of individuals from the different sectors of the society that often presents a problem in communication. The transfer of new research findings to the users is now becoming a slow, tedious and costly process.

One of the practical objectives of this series on *Engineering Application of Fracture Mechanics* is to provide a vehicle for presenting the experience of real situations by those who have been involved in applying the basic knowledge of fracture mechanics in practice. It is time that the subject should be presented in a systematic way to the practising engineers as well as to the students in universities at least to all those who are likely to bear a responsibility for safe and economic design. Even though the current theory of linear elastic fracture mechanics (LEFM) is limited to brittle fracture behavior, it has already provided a remarkable improvement over the conventional methods not accounting for initial defects that are inevitably present in all materials and

G. C. Sih and E. E. Gdoutos (eds): Mechanics and Physics of Energy Density, IX–XI.
© 1992 Kluwer Academic Publishers. Printed in the Netherlands.

structures. The potential of the fracture mechanics technology, however, has not been fully recognized. There remains much to be done in constructing a quantitative theory of material damage that can reliably translate small specimen data to the design of large size structural components. The work of the physical metallurgists and the fracture mechanicians should also be brought together by reconciling the details of the material microstructure with the assumed continua of the computational methods. It is with the aim of developing a wider appreciation of the fracture mechanics technology applied to the design of engineering structures such as aircrafts, ships, bridges, pavements, pressure vessels, off-shore structures, pipelines, etc. that this series is being developed.

Undoubtedly, the successful application of any technology must rely on the soundness of the underlying basic concepts and mathematical models and how they reconcile with each other. This goal has been accomplished to a large extent by the book series on *Mechanics of Fracture* started in 1972. The seven published volumes offer a wealth of information on the effects of defects or cracks in cylindrical bars, thin and thick plates, shells, composites and solids in three dimensions. Both static and dynamic loads are considered. Each volume contains an introductory chapter that illustrates how the strain energy criterion can be used to analyze the combined influence of defect size, component geometry and size, loading, material properties, etc. The criterion is particularly effective for treating mixed mode fracture where the crack propagates in a nonself similar fashion. One of the major difficulties that continuously perplex the practitioners in fracture mechanics is the selection of an appropriate fracture criterion without which no reliable prediction of failure could be made. This requires much discernment, judgement and experience. General conclusion based on the agreement of theory and experiment for a limited number of physical phenomena should be avoided.

Looking into the future the rapid advancement of modern technology will require more sophisticated concepts in design. The micro-chips used widely in electronics and advanced composites developed for aerospace applications are just some of the more well-known examples. The more efficient use of materials in previously unexperienced environments is no doubt needed. Fracture mechanics should be extended beyond the range of LEFM. To be better understood is the entire process of material damage that includes crack initiation, slow growth and eventual termination by fast crack propagation. Material behavior characterized from the uniaxial tensile tests must be related to more complicated stress states. These difficulties should be overcome by unifying metallurgical and fracture mechanics studies, particularly in assessing the results with consistency.

This series is therefore offered to emphasize the applications of fracture mechanics technology that could be employed to assure the safe behavior of engineering products and structures. Unexpected failures may or may not be critical in themselves but they can often be annoying, time-wasting and discrediting of the technical community.

Bethlehem,   G.C. SIH
Pennsylvania   *Editor-in-Chief*

Foreword

This volume on the application of energy density in mechanics and physics is concerned with the field of material/structure technology. Theoretical and experimental contributions were presented as lectures at the Democritus University of Thrace, Xanthi, Greece, during the period of July 17 to 20, 1989. They included individuals from Australia, Canada, German Democratic Republic, Greece, Hungary, Italy, Poland, People's Republic of China, USA und USSR. Most rewarding are the exchange of new ideas and informal discussions that were made possible only by a limited size audience. Discussions on the development of new materials and their applications were particularly relevant to the on-going research of the European Common Community.

The co-organizers of this symposium are indebted to those who have spent many hours of their valuable time in preparing the manuscript. Financial support from

Ministry of Industry, Research and Technology
Democritus University of Thrace
Lehigh University
Prefecture of Xanthi
Municipality of Xanthi
Commercial Bank of Greece
National Bank of Greece
Cooperative Cigarette Company (SEKAP)
Cooperative Meat Company (SEPEK)
Cooperative Milk Company

are also gratefully acknowledged. The following members of the local organizing committee are credited with operating the visual equipment and organizing social activities:

Dr. E.E. Gdoutos
Dr. D.A. Zacharopoulos
Dr. G. Papakaliatakis
Mrs. Z. Adamidou
Mr. A. Betzakoglou
Ms. C. Gemenetzidou
Mr. N. Prassos

Xanthi, Greece   G.C. SIH
July 1989   E.E. GDOUTOS
                                                                    Co-Organizers

Editors' Preface

The motivation to publish this volume stems from the central role that energy density plays in the formulation of continuum mechanics theories and characterization of material/structure behavior. So many developments in mechanics and physics depended on the positive definiteness of volume energy density and its scalar invariant property. Examples are the Kirchhoff uniqueness proof in the linear theory of elasticity and determination of constitutive relations based on the theoretic invariant approach. Proposed by E. Beltrami as early as 1903 was that the limiting capacity of a solid could be related to elastic energy stored in a unit volume of material. This unique character of the volume energy density function was also recognized in the 1960's by the late L. Gillemot as a useful parameter for characterizing the plastic behavior of metals and weldments.

The versatility of this quantity has been expanded over the years to account for failure by yielding and fracture which included initiation, stable growth and onset of global instability. It matters not of material type nor of the source of energy whether chemical, electrical or otherwise. Disruptions in material/ structure behavior can, in general, be associated with the fluctuations of energy in a unit volume and/or surface of material. What is not commonly recognized is the relation between surface and volume energy density even though it has been explicitly expressed in the crystal nucleation theory of Gibb where the energy proportional to the nucleus volume exchanges directly with that to the nucleus surface area. Further clarification on how energy density could be applied to formulate theories and to explain physical phenomena is one of the main objectives of the fourteen chapters in this volume.

In their work, O.A. Bukharin and L.V. Nikitin derived the volume energy density for a linear viscoelastic solid and showed that the threshold stresses corresponding to the initiation of stable and unstable crack growth can be

G. C. Sih and E. E. Gdoutos (eds): Mechanics and Physics of Energy Density, XV–XVIII.
© 1992 *Kluwer Academic Publishers. Printed in the Netherlands.*

obtained in a straightforward manner bypassing much of the complexities required in previous works on the same subject. M.M. Smith and G.F. Smith made use of the scalar property of the strain energy density function for generating the invariants that describe the material symmetry. Developed are procedures for deriving constitutive relations. Use is also made of the strain energy density by A.A. Liolios to establish upper and lower bounds in the numerical analysis of problems in elastostatics and elastodynamics.

A dislocation model was proposed by W. Lung, L.Y. Xiong and S. Liu to obtain the fracture behavior of macrocrack under the influence of plasticity. Influence of the ligament from the crack to the plate edge on plasticity can be related to the change in volume with surface, a quantity that interconnects the surface and volume energy density. Microscopic damage and critical increment of crack growth prior to macroinstability for thermoplastics were successfully explained by T. Vu-Khanh, B. Fisa and J. Denault. They applied the strain energy density criterion and showed that the addition of fillers led to a reduction in the critical strain energy density. This can be interpreted as an enhancement of the resistance of thermoplastics to fracture. V.S. Ivanova modelled the microdamage and macrodamage of metal alloys by considering the critical distortional and dilatational energy density. The former was identified with the latent heat of melting and latter with the internal energy density calculated from the heat capacity at constant pressure. Analytical results are related to damage zone sizes at the microscopic and macroscopic scale level; they compared well with experiments. As the density of microcracks in a unit volume of material changes, the corresponding energy density would be affected as discussed by N. Laws. The response for open and closed cracks can be different and the influence on the strain energy density function must be determined accordingly.

The energy density criterion is employed by J. Lovas to evaluate the damage in aluminum and stainless steel bars that are extruded through dies of different shapes. Sites of potential failure by fracture are predicted from the maximum of minimum strain energy density. An analytical/experimental procedure was used and the predictions agreed with the tests. Presented by K.F. Fischer is how mixed mode fracture could be characterized using the criterion of strain energy density. The direction of crack initiation is assumed to coincide with the location of minimum energy density factor or S^{mi}_n while the onset of crack growth is assumed to occur when S^{min} reaches some critical value. Results are displayed in terms of plots of Mode I and II stress intensity factors and discussed in connection with other failure criteria. Critical loads to initiate crack growth and fracture trajectories for cracks at interface of inclusion and matrix are obtained by E.E. Gdoutos and M.A. Kattis. They made consistent use of the strain energy density criterion. The influence on change in the local material behavior on subcritical crack growth was examined by G.E. Papakaliatakis, E.E. Gdoutos and E. Tzanaki. Each material element along the prospective path of crack growth can follow a different elastic-plastic stress/strain response and its behavior can be affected by the previous loading step. Crack growth is assumed to occur in segments corresponding to a constant ratio of the strain energy density factor and growth segment. Strain rate and strain rate history dependency led to crack growth data that deviated appreciably from those found by the classical theory of plasticity.

As the failure behavior of small specimens or structures can differ significantly from the larger ones, A. Carpinteri applied the strain energy density criterion to obtain the scaling relations that include the combined influence of specimen size, material and loading steps. Crack growth resistance curves are found for concrete-like materials that undergo hardening and softening. For a number of years, R. Jones, M. Heller and L. Molent at the Aeronautical Research Laboratory in Melbourne, Australia, have effectively applied the strain energy density theory to the design of reinforcement for repairing aircraft skin structures damaged by stress corrosion cracking. Summarized in their Chapter are some of the more recent works that illustrate how the temperature field can be measured and used to determine the energy density field of flawed structures.

Perhaps, one of the most significant post World War II developments in the field of material/structure behavior is the recognition that design requirements based on classical continuum mechanics solutions alone were not sufficient. Additional failure criteria must be supplemented to account for the criticality of defects or discontinuities. The impetus were thus provided to put the 1921 concept of A.A. Griffith into practice. A quantity that became the focus of attention in fracture mechanics research was the energy required to create a unit area of fracture surface, the equivalent of which are many only for the idealized situation of no energy dissipation where the material can recover completely. The restrictive character of linear elastic fracture mechanics gave information only on the terminal condition which does not provide the safety measures required in many of the high performance structures. Needed is the assessment of how and where defects would initiate in addition to their subsequent growth behavior. To this end, the volume energy density criterion were advanced to address the complete process of defect initiation, subsequent growth and onset of instability. Much of the work can be found in the introductory chapters of the seven volumes of the series on Mechanics of Fracture and the references therein. It represented a contrast to the surface energy density quantity associated with energy release rate, stress intensity factor, crack opening displacement, etc. Failure analyses in fracture mechanics have thus followed two separate schools of thought: one on surface energy density and the other on volume energy density. This trend has prevailed up to this date. Discussion on energy density would be incomplete without touching on the reconciliation of the two approaches. This can be made possible only by revoking a fundamental assumption in continuum mechanics. That is by not letting the size of the continuum element to vanish in the limit. Because it is the *finite* rate change of volume with surface that links the surface and volume energy density. The ways with which this can be done is summarized in the first chapter of this volume. It hinges on interlacing the mechanical and thermal effects so that the complete hierarchy of the material damage process can be assessed quantitatively without ambiguity. Deformation and temperature changes in uniaxial specimens, once synchronized, are found to be highly nonhomogeneous. The corresponding thermal/mechanical properties are transitory in character. Derived are nonequilibrium results that correlate well with real time data. Unless displacement and/or temperature data are quantified by the time and size over which measurements are made, they would be vulnerable to miscue.

To conclude, the editors wish to accredit the authors for their consulted efforts that made the material coherent as a volume. The knowledge gained in this exposition hopefully would simulate others to further the use of energy density in other fields of physics and engineering. Completion of the manuscripts depended much on the assistance of Mrs. Barbara DeLazaro and Constance Weaver whose efforts are acknowledged.

Xanthi, Greece   G.C. SIH
July 1989   E.E. GDOUTOS
                                                                        Volume Editors

Contributing authors

O.A. Bukharin
USSR Academy of Sciences, Moscow, USSR

A. Carpinteri
Politecnico di Torino, Torino, Italy

J. Denault
National Research Council Canada, Boucherville, Quebec, Canada

B. Fisa
National Research Council Canada, Boucherville, Quebec, Canada

K.-F. Fischer
Ingenieurhochschule Zwickau, Germany

E.E. Gdoutos
Democritus University of Thrace, Xanthi, Greece

M. Heller
Aeronautical Research Laboratory, Melbourne, Australia

V.S. Ivanova
USSR Academy of Sciences, Moscow, USSR

R. Jones
Aeronautical Research Laboratory, Melbourne, Australia

M.A. Kattis
Democritus University of Thrace, Xanthi, Greece

N. Laws
University of Pittsburgh, Pittsburgh, Pennsylvania

A.A. Liolios
Democritus University of Thrace, Xanthi, Greece

S. Liu
Institute of Metal Research Academia Sinica, Shenyang, China

J. Lovas
Technical University of Budapest, Budapest, Hungary

C.W. Lung
Institute of Metal Research Academia Sinica, Shenyang, China

L. Molent
Aeronautical Research Laboratory, Melbourne, Australia

L.V. Nikitin
USSR Academy of Sciences, Moscow, USSR

G.E. Papakaliatakis
Democritus University of Thrace, Xanthi, Greece

G.C. Sih
Lehigh University, Bethlehem, Pennsylvania

G.F. Smith
Lehigh University, Bethlehem, Pennsylvania

M.M. Smith
Lehigh University, Bethlehem, Pennsylvania

E. Tzanaki
Democritus University of Thrace, Xanthi, Greece

T. Vu-Khanh
National Research Council Canada, Boucherville, Quebec, Canada

L.Y. Xiong
Institute of Metal Research Academia Sinica, Shenyang, China

At the banquet: speakers, organizers, publishers and guests.

Committee of the ladies.

Greetings from the Mayor of Xanthi, K. Benis, at Town Hall.

Outing to the Island of Thassos.

At the footsteps of Xenia Hotel.

Chatting with the Rector E. Galousis of Democritus University of Thrace.

Synchronization of thermal and mechanical disturbances in uniaxial specimens

1.1 Introductory remarks

Nonequilibrium phenomena in solids are characterized by the continuous change of thermal/mechanical state variables with location and time. The process can be highly nonhomogeneous and transient. Assessment of non-equilibrium processes are, as a rule, bypassed not by choice but by the lack of a suitable theory that could collate "real time" measurements which are distinguished from those averaged over a period of time whether statistically or otherwise. While nonequilibrium statistical mechanics delves into the details of particle interaction and trajectory, it becomes impossible to manage, not only in computation but also in principle, when applied to determine macroscopic effects. The inexact modelling of crystal atoms and gas atoms as particles introduces approximations of unknown nature into the determination of their trajectories. Inadequately treated in the statistical models is also the temperature changes caused by the heat generated from the motion of the mutually interacting particles.

When a tensile bar is stretched regardless of the rate of deformation, its thermal/mechanical state variables change with time and vary from location to location. More specifically, the stress and strain response and temperature become spatially nonuniform. What occurs intrinsically is the exchange of thermal and mechanical energy. A net flow of heat prevails locally making the process one of nonequilibrium manifested by a lowering of the temperature below ambient at the initial stage. Heat is transferred from the surrounding into the specimen before the opposite happens. This reversal of heat flow is a macroscopic phenomenon and would correspond to a negative entropy in classical equilibrium thermodynamics. Contrary to the *a-priori* judgment in continuum mechanics theories that the onset of temperature rise occurs at the yield point,

G. C. Sih and E. E. Gdoutos (eds): Mechanics and Physics of Energy Density, 1–34.
© 1992 *Kluwer Academic Publishers. Printed in the Netherlands.*

experiments have shown otherwise [1–6]. For the most part of the uniaxial stress/strain history, the temperature remained below ambient passing well beyond yield. A considerable increase in temperature takes place as the specimen approaches the breaking point where ultimate stress and strain are recorded. Refer to the recent theory and experiment for steel [1] and aluminum [2]. Explanations offered by the equilibrium theories [3–6] on cooling/heating are suspect on fundamental grounds because they are valid only for very large values of time until the thermal/mechanical state variables are stabilized. The cooling would have by then disappeared.

One of the objectives of this work is to show that the time response and element size chosen for analysis are not unrelated. They cannot be selected arbitrarily. Macroscopic time/size scale is distinctly different from microscopic time/size scale [7]. Unless such a distinction is made, thermal and mechanical disturbances would not be synchronized. Experimental data on cooling/ heating of uniaxial specimens will be discussed to illustrate the highly nonequilibrating character of the response. Material properties are shown to be spatial and load history dependent which cannot be resolved by appealing for a more general form of the constitutive relation.

1.2 System inhomogeneity and continuity

Thermodynamically speaking, physical systems would by nature interact with their surroundings such that their states change constantly. Whether the deviations from the equilibrium state are important or not depends on the specific problem at hand. In an "open" system, the conditions on the boundary change with the surrounding and should be derived. This is in contrast to a "closed" system where the boundary conditions are imposed. Isothermal and adiabatic are two of the commonly assumed conditions. Their application in field theories for establishing the existence of potential functions such as the strain energy density function in the classical theory of elasticity [8] has relied on arguing for a slowly or suddenly applied load. Despite the long historical account on research concerning the measurement of physical constants under isothermal and adiabatic conditions, the justification for applying the open system data obtained from uniaxial test specimens to theories such as elasticity, viscoelasticity, etc., based on closed-system assumptions were never given. Only in recent times has the thermal/mechanical state of affairs for a uniaxial tensile specimen been investigated [9, 10]. This was accomplished

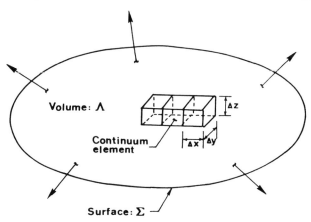

Figure 1.1. Schematic of a continuum divided into finite numbers of elements.

by solving a two-phase solid/gas interaction problem. For a 6061–T6 aluminum material stretched at a displacement rate of 1.27×10^{-4} cm/sec, the thermal disturbances created by the slowly stretched solid sustained for a distance more than half the specimen length before they died out. The conditions across the solid/gas interface were neither isothermal nor adiabatic. The specimen temperature remained below ambient for several minutes. The local stress, strain, temperature and energy density vary from one part of the system to another at each time instance. The conventional assumption that the state of affairs remains constant over a certain gage length cannot be justified. Equilibrium theories would not yield such predictions because the thermal and mechanical disturbances would not be synchronized.

Nonhomogeneous behavior. Aside from changes arising from the material microstructure, inhomogeneity can prevail on account of nonuniform exchange of thermal and mechanical energy in a unit volume of material. This can be illustrated schematically in Figure 1.1 for a three-dimensional body. Let the volume Λ be bounded by the surface Σ in the coordinate system (x, y, z). The body is subdivided into a set of small elements with dimensions Δx, Δy and Δz. Each element is sufficiently small such that its behavior could be described by the local value of u_i and Θ. The temperature is Θ and displacements are u_i; they are one of the same operation and should be interlaced in the formulation of continuum mechanics theories. Some *a-priori* requirements on the size of $(\Delta x, \Delta y, \Delta z)$ and time rate of change of u_i and Θ should be set as

- Element size should be made sufficiently small such that variations of u_i and across $(\Delta x, \Delta y, \Delta z)$ are negligibly small.

- The time rate change of u_i and Θ must also be considered in the selection of element size. This is because the time required to achieve equilibration or uniformity decreases with element size.

Since the subdivision of body into elements is conceptual, there is no limitation to the smallest element or fastest time that could be made. No difficulties arise in relating atomic, microscopic and macroscopic events if scaling of size/time/temperature is considered [7]. The requirement that each micro or macro element must contain a sufficient number of atoms in order to preserve continuity, may be circumvent. Uniqueness of each atom or particle, however minute, could be identified by the local surface and/or volume energy density,* provided that the system is modelled with the following properties:

- The conceptual elements in the system must be finite and need not be of uniform size. They would, in fact, be different in size because of inhomogeneity and must satisfy kinematic compatibility.
- Element size must be time and temperature-dependent in addition to deformation.

A theory with the aforementioned capability imposes no limit on the relative fluctuations of the local material properties and prescribes the resolution of observation on a continuous scale level. The isoenergy density theory presented in [11–13] possesses such a capability.

Continuity requirement. All continuum theories, when applied to characterize real systems, should self-guide against the violation of material continuity. Analysis of the stress and strain in an initially smooth uniaxial specimen, for example, should be interrupted before reaching the ultimate stress and strain point, because material continuity would be violated by the creation of a discontinuity prior to that time. A decision would have to be made on the size of the discontinuity, say 10^{-2} cm or 10^{-1} cm in linear dimension. As the damage process is cumulative and time dependent, thresholds of both surface and volume energy density denoted, respectively, by $(dW/dA)_c$ and $(dW/dV)_c$ should be established; they would change with time because material degrades gradually.

Suppose that dW/dA and dW/dV represent the time dependent surface and volume energy density in an element $\Delta x\, \Delta y\, \Delta z$ in Figure 1.1 as the body is subjected to increasing strains. During the initial stage of load transfer through Σ, the surface energy density dW/dA would be high, while the volume

---

* In the isoenergy density theory [11–13], the energy density quantities depend only on the displacements. No *a-priori* assumption is made on the constitutive relation and hence dW/dA and/or dW/dV are determined without a knowledge of the stresses.

energy density dW/dV would be low. As more and more energy is transmitted into the region Λ, the increase in dW/dV would catch up to that in dW/dA. These features are shown schematically in Figures 1.2 and 1.3. It is seen that dW/dA rises rapidly for small time and then levels off, which is in contrast to dW/dV which increases slowly at first and then rapidly for large time. The following possibilities could arise

$$\frac{dW}{dA} > \left(\frac{dW}{dA}\right)_c; \qquad \frac{dW}{dV} \simeq \left(\frac{dW}{dV}\right)_c \quad \text{or} \quad \frac{dW}{dV} < \left(\frac{dW}{dV}\right)_c \qquad (1.1)$$

and

$$\frac{dW}{dV} > \left(\frac{dW}{dV}\right)_c; \qquad \frac{dW}{dA} \simeq \left(\frac{dW}{dA}\right)_c \quad \text{or} \quad \frac{dW}{dA} < \left(\frac{dW}{dA}\right)_c. \qquad (1.2)$$

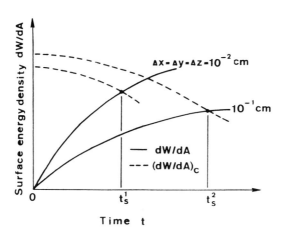

Figure 1.2. Failure time associated with element size for surface energy density.

Equation (1.1) implies that at a given location, dW/dA, has reached critical, $(dW/dA)_c$, and failure ensues. Interaction of time and element size at failure stated in equation (1.1) is given in Figure 1.2.

Creation of new surface area of the order of 10^{-2} cm in linear dimension is predicted at t_s^1. Failure by dW/dV reaching critical would not occur at this stage because $t_s^1 < t_v^1$. Refer to Figure 1.3 for the definition of t_v^1. Should this failure be assumed to have negligible effect on continuity, the morphology of the system would then be retained and the analysis continues. A larger surface area with a linear dimension of 10^{-1} cm would be predicted at a later time $t_s^2 > t_s^1$. Whether failure at t_s^2 would occur or not depends on whether the volume

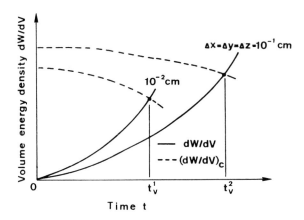

Figure 1.3. Failure time associated with element size for volume energy density.

energy density dW/dV has reached the condition in equation (1.2) or not prior to t_s^2. That is, if t_v^2 in Figure 1.3 is smaller than t_s^2 in Figure 1.2, then a unit volume of material would first fragment with a linear dimension of 10^{-1} cm. This might be enough to interrupt continuity, particularly if such a condition has spread over many elements over a region. This region must then be excluded from the analysis to restore continuity of the region Λ, excluding the discontinuity.

Adjustment in material properties outside the locally failed region should also be made because it takes into account the effect of damage accumulation. In this way, the hierarchy of the material damage process could be uniquely and consistently characterized, starting from the atomic to the macroscopic level. The initial state of the crystalline microstructure for a metal [13] with varying properties from grain to grain could be specified by the fluctuations in the surface and volume energy density. The same applies to composites with initial inhomogeneous and anisotropic material properties.

1.3 Simultaneity of displacement and temperature change

Ordinary thermocouples do not possess the time response required for measuring the change in temperature that is compatible with that in displacement or strain. This is one of the main reasons why the cooling/heating behavior of uniaxial specimens did not receive the attention it deserves until the advent of high-performance semiconductors that upgraded the time response in temperature measurements.

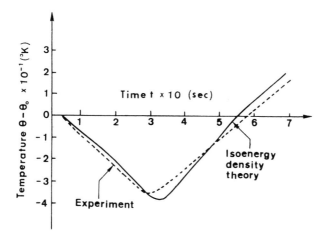

Figure 1.4. Nonequilibrium cooling/heating response for uniaxially stretched SAFC-4OR specimen at 0.2 cm/min.

Earlier temperature measurements on mild steel [3] attributed jumps in temperature to the onset of yielding is suspect in view of the more recent data on cooling/heating [1, 2].

Uniaxial tension of SAFC-40R steel. Temperature changes $\Theta - \Theta_0$ in a SAFC-40R steel specimen stretched at a rate of 0.2 cm/min were obtained in [4] and the results are represented by the dotted curve in Figure 1.4. Note that cooling occurred for more than 30 sec while the specimen remained below ambient Θ_0 for almost 60 sec. Levelling of the stress/strain occurred after 30 sec. No association of the knee in the temperature curve with the so-called "yield point" on the stress/strain curve could be made,* simply because the concept of yielding belongs to the mathematical theory of plasticity that invokes equilibrium. Predicted by the isoenergy density theory [1] is the solid curve in Figure 1.4 where the agreement with experiment is good. Dissipation energy is found to be significant, even when the stress and strain relation were nearly linear in the time range up to 20 sec. Linearity does not imply reversibility.

Torsion of 6061-T6 tube. The cooling/heating phenomenon occurs not only for the stretching of metals and polymers, but it also prevails in specimens under torsion [14]. Experiments on Alcoa L seamless 6061-T6 aluminum tubes with dimensions $0.625 \times 0.058 \times 9.5$ in were twisted at 10 degrees per minute. Table 1.1 gives the numerical values of the temperature change as a function of time for two specimens: one with $\Theta_0 = 27.5°C$ and the

---

* The coincidence of yield point obtained by the 0.2% offset with the trough in the temperature curve is purely coincidental [3]. No physical significance could be attached to it.

other with $\Theta_0 = 28°C$. The time elapsed when cooling terminates and heating begins is significant. This is displayed graphically in Figure 1.5. Such a phenomenon was disclaimed* by the thermoviscoplasticity theory in [15].

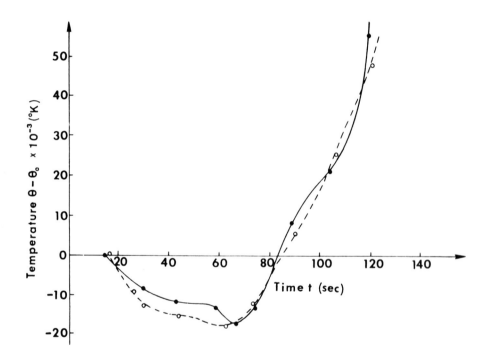

Figure 1.5. Time-history of temperature for Alcoa L-seamless tube 6061-T6 (0.625 × 0.058) in torsion at 10°/min.

---

*   The result on cooling and heating for the uniaxial tensile specimen in [6] was contrived because the invokement of equilibrium in the thermoviscoplasticity theory would exclude the reversal of heat transfer between the system and its surroundings.

Table 1.1. Time history of surface temperature for 6061-T6 aluminum tube under torsion.

	Time t (sec)	Temperature change $(\Theta \text{-} \Theta_0) \times 10^{-3}$ (°K)
Dotted curve	0	0
($\Theta_0 = 27.5$°C)	0	15
	− 9.17	30
	−11.00	45
	−11.92	60
	−16.51	67.5
	−13.76	75
	9.16	90
	20.15	105
	54.86	120
	82.18	135
Solid curve	0	16
($\Theta_0 = 28.0$°C)	− 9.13	27
	−13.05	31
	−15.22	45
	−16.96	63
	−12.18	75
	4.34	90
	26.03	105
	47.65	120

1.4 Isoenergy density theory

Formulation of macroscopic nonequilibrium theories can be made by not letting the element size to vanish such that the rate change of volume with surface is a finite quantity. Referring to the three orthogonal planes $i = x, y, z$ in Figure 1.6, the three components $(\Delta V / \Delta A)_i$ are shown on the tetrahedron where **n** is the outward unit normal vector. As the volume element ΔV or

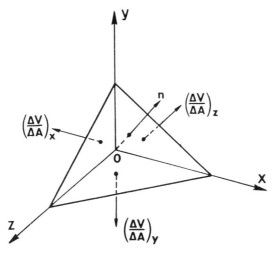

Figure 1.6. Components of rate change of volume with surface referred to a tetrahedron.

$\Delta x \Delta y \Delta z$ remains finite, the surface tractions $\overset{n}{T_i}$ and body forces are no longer independent:

$$\overset{n}{T_i} = \sigma_{ji}\, n_j + \rho\,(\ddot{u}_i - h_i)\left(\frac{\Delta V}{\Delta A}\right)_n,\tag{1.3}$$

where σ_{ij} are the stress components, $\ddot{u}_i$ are the accelerations and h_i the body force components with ρ being the mass density. The boundary conditions

$$\overset{n}{T_i} = \sigma_{ji}\, n_j\tag{1.4}$$

used in classical field theories are recovered from equation (1.3) only when $(\Delta V/\Delta A)_n \to 0$. The other choice would be if $\rho(\ddot{u}_i - h_i) \to 0$ or the product is negligible in comparison with $|\sigma_{ij}|$. The inertia term $\rho\ddot{u}_i$ for a local element may not be negligible, even though the global behavior could be regarded as one of static loading. On physical grounds, a finite time step of loading is involved, however small, in the numerical analysis for which $\rho\ddot{u}_i$ remains finite. The number of steps affects the kinematics of the finite element analysis and, hence the outcome. Discrepancy between equations (1.3) and (1.4) may not be negligible near a free surface or interface. It can always be checked by computing for $(\Delta V/\Delta A)_i$ from the displacements. Large deviations in $(\Delta V/\Delta A)_i$ could be interpreted either as incompatibility between the continuum field theory and discrete character of numerical analysis or as inadequacy of the classical field theory for having neglected the interaction of surface and volume energy that will be discussed subsequently.

Interaction of surface and volume energy. Surface is conceived mathematically to define volume. A basic assumption in the formulation of field theories is that a continuum, say with volume Λ enclosed by surface Σ, can be divided and subdivided indefinitely into any number of finite elements with volume ΔV or $\Delta x\, \Delta y\, \Delta z$ as discussed earlier in connection with Figure 1.1. Each surface of the element can be identified with an area ΔA_i $(i = x,\ y,\ z)$ which changes simultaneously with ΔV and, hence the quantity $(\Delta V/\Delta A)_i$. If $(\Delta W/\Delta A)_i$ is used to denote the energy associated with a unit area of the continuum element, then $\Delta W/\Delta V$ is the corresponding energy associated with a unit volume of the same element. They are related as [11–13]:

$$\left(\frac{\Delta W}{\Delta A}\right)_i = \left(\frac{\Delta V}{\Delta A}\right)_i \frac{\Delta W}{\Delta V},\quad i = x,\ y,\ z.\tag{1.5}$$

This exchange of surface and volume energy is not included in classical continuum mechanics theories because of the assumption $\Delta V/\Delta A \to 0$.

Suppose that the differential change of the volume energy density dW/dV can be determined in a one-dimensional space of τ and e by the integral

$$\frac{dW}{dV} = \int \tau(e) \, de \tag{1.6}$$

in which e may stand for any one of the nine displacement gradients

$$e_{ij} = \frac{\partial u_i}{\partial x_j}, \quad i, j = x, y, z. \tag{1.7}$$

If the function $\tau(e)$ can further be expressed as

$$\tau = \int \lambda \frac{dV}{dA} \, de, \tag{1.8}$$

then equation (1.6) can be written as

$$\frac{dW}{dV} = \int\int \lambda \frac{dV}{dA} \, de \, de. \tag{1.9}$$

Since dV/dA depends only on e and λ is assigned to denote a particular state of energy, dW/dV can be found with a knowledge of the displacement field or its gradient. The same applies to $(dW/dA)_i$ in view of equation (1.5). Extension of the one-dimensional form of equation (1.6) to multidimensions can thus be done. More specifically, it would be useful to establish the condition

$$\mathscr{W}(\tau, e) \to \mathscr{W}(\tau_{ij}, e_{ij}) \quad \text{in } \Lambda, \tag{1.10}$$

where both τ_{ij} and e_{ij} are components of the second-order tensor. The notation $\mathscr{W}$ for the volume energy density dW/dV has been introduced.

Isoenergy surface. An isoenergy surface possesses the property that the transmission of energy on such a surface is directional dependent. The position of such a surface is different for each element whose size depends on the local rotation and deformation that operate as a single process. If ξ_i $(i = 1, 2, 3)$ are the current state coordinates of an element as distinguished from the reference coordinates x_i $(i = 1, 2, 3)$, then the condition

$$\mathscr{S}_1 = \mathscr{S}_2 = \mathscr{S}_3 \to \mathscr{S} \tag{1.11}$$

determines the orientation of the element for which the same $\mathscr{S}$ prevails on all orthogonal surfaces. In equation (1.11), $\mathscr{S}_{\cdot}$ are the surface energy densities $(dW/dA)_i$ in equation (1.5). In this respect, τ and e can be referred to, respectively, as the isostress and isostrain. Moreover, e can be shown to represent any one of the nine displacement gradients

$$e_{ij} = \frac{\partial u_i}{\partial \xi_j}, \quad i, j = 1, 2, 3.$$ (1.12)

The same applies to τ which is paired with e or τ_{ij} with e_{ij}. In view of equations (1.5) and (1.11), it is easily seen that

$$\mathcal{V}_1 = \mathcal{V}_2 = \mathcal{V}_3 \rightarrow \mathcal{V},$$ (1.13)

where $\mathcal{V}$ denotes dV/dA. The rate change of volume with surface on the isoenergy surface is also the same in all directions.

Thermal/mechanical interaction. An important feature of the isoenergy density theory [11–13] is that the motion of mass elements and resulting temperature change are synchronized. Interdependence of these two effects cannot be assumed but must be determined as

$$\frac{\Delta\Theta}{\Theta} = -\lambda\mathcal{V}\frac{\Delta e}{\Delta\mathcal{D}/\Delta e}$$ (1.14)

where Θ represents the nonequilibrium temperature in contrast to T in classical thermodynamics that applies only to equilibrium states. The dissipation energy density $\mathcal{D}$ is determined on the isoenergy density surface; it is a positive definite quantity

$$d\mathcal{D} = d\mathcal{W} - d\mathcal{A}, \quad \mathcal{D} \geqslant 0; \quad \frac{d\mathcal{D}}{dt} \geqslant 0$$ (1.15)

with $\mathcal{A}$ being the available energy density. Equation (1.15) is equivalent to the conservation of energy. The dissipated and available energy are mutually exclusive and, hence, there is no loss in generality in the linear dependence state in equation (1.15). Because Θ can be determined from $\mathcal{V}$, e and $\mathcal{D}$ without the concept of heat, the isoenergy density formulation represents a fundamental departure from thermodynamics. Irreversibility is reflected through the $\mathcal{H}$-function:

$$d\mathcal{H} = -\frac{d\mathcal{D}}{\Theta}.$$ (1.16)

The negative sign indicates work done on the system. The change of $\mathcal{H}$ can be positive, zero or negative. It is not related to the Boltzmann H-theorem in statistical dynamics for dilute gases.

Isostress and isostrain. Since the isoenergy density state is completely defined by $(\mathcal{V}, e)$ as shown by the form of $\mathcal{W}$ in equation (1.9), there is really no need for introducing the concept of stress. For the sake of familiarity, however, an isostress quantity τ referred to the isoenergy density surface will be used. Refer to equation (1.8). A one-to-one correspondence between $(\mathscr{V}, e)$ and (τ, e) can be established without loss in generality as given by equation (1.8). A unique feature of the isoenergy density theory [11–13] can be stated as follows:

The state of an isoenergy element defined by $(\mathscr{V}, e)$ or (τ, e) is determined for each load step and time increment without a preknowledge of the so-referred to "constitutive relations" in classical field theory.

Only the relation between τ and e or $\mathscr{V}$ and e for the reference state of a material is needed. Once the loading steps are specified, adjustment on (τ, e) is made for each of the subsequent steps. In other words, the reference state from 0 to 1 in Figure 1.7 or $(\mathscr{V}^0, e^0)$ is assigned while the subsequent steps $(\mathscr{V}^1, e^1), (\mathscr{V}^2, e^2)$, etc., are determined, the loci of which gives the response of a typical element. It is a self-adaptive procedure that accounts for the system inhomogeneity.

Governing equations. As in the classical field theories, the governing equations can be derived by application of the conservation of linear and angular momentum and conservation of energy. Nonequilibrium, nonlocal, irreversible, and large/finite deformation effects are all included in addition to having finite element size. This necessitates the distinction between the Cauchy stress $\boldsymbol{\sigma}$ and isostress $\boldsymbol{\tau}$ which are related as

$$\tau_{ji} = \sigma_{ji} + \rho(\ddot{u}_i - h_i) \mathscr{V}_{,j}. \tag{1.17}$$

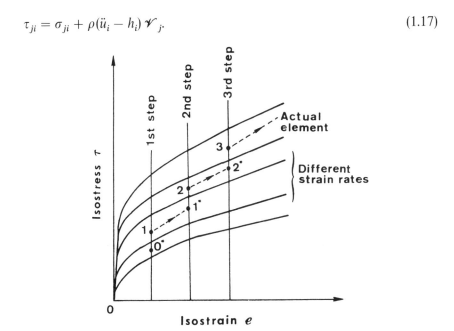

Figure 1.7. Determination of stress and strain response in steps on isoenergy density plane.

The isostress tensor τ is not symmetric on account of the nonlocal character of the deformation. Making use of equation (1.17), the conditions in equation (1.3) can be written as

$$\overset{n}{T_i} = \tau_{ji} n_j, \quad \tau_{ij} \neq \tau_{ji}. \tag{1.18}$$

The conservation of energy applied to the system in Figure 1.1 yields

$$\int_\Lambda \rho h_i \dot{u}_i \, dV + \int_\Sigma \overset{n}{T_i} \dot{u}_i \, dA = \int_\Lambda \rho \ddot{u}_i \dot{u}_i \, dV + \frac{d}{dt} \int_\Lambda \mathscr{W} \, dV \tag{1.19}$$

in which

$$\frac{d\mathscr{W}}{dt} = \tau_{ij} \dot{e}_{ij}. \tag{1.20}$$

Note that $\mathscr{W}$ is not an elastic potential. It applies to an irreversible and dissipative process as implied by equations (1.15) and (1.16). The equations of equilibrium are

$$\frac{\partial \tau_{ji}}{\partial \xi_j} + \rho h_i = \rho \ddot{u}_i. \tag{1.21}$$

Existence and boundness. The existence of the isoenergy density function $\mathscr{W}$ has been proved in [12]. Only the theorem will be stated as follows:

There exists an isoenergy density function $\mathscr{W}$ that can be obtained by integrating $\lambda \mathscr{V}$ twice with respect to any one of the nine displacement gradients $\partial u_i / \partial \xi_j$ (i, j = 1, 2, 3) where $\mathscr{V}$ is the change of volume with surface for a given λ.

Mathematical equivalence of the foregoing statement is that

$$\mathscr{W} = \int \tau_{11} \, de_{11} = \int \tau_{12} \, de_{12} = \cdots = \int \tau_{23} \, de_{23} \to \int \tau \, de. \tag{1.22}$$

This means that any one of the nine pairs $(\tau_{11}, e_{11}), (\tau_{12}, e_{12}), ..., (\tau_{23}, e_{23})$ or (τ, e) can be used to yield the same $\mathscr{W}$. Hence, the uniaxial isostress τ versus the isostrain e plot provides a general description of the energy state, even though the other stress components τ_{ij} are also acting on the element.

Because nonequilibrium implies nonuniqueness, uniqueness proof can be provided only for equilibrium states [12]:

Isoenergy equilibrium states $(\mathscr{V}, e)$ or (τ, e) are unique for positive definite isoenergy density function $\mathscr{W}$, and the satisfaction of equilibrium and continuity.

The nonequilibrium fluctuation of isoenergy states can only be bounded by their neighboring equilibrium states. Boundness and limit have been rigorously established and can be summarized as [12]:

Fluctuations of nonequilibrium/irreversible isoenergy states are bounded by equilibrium/irreversible isoenergy states and they tend to definite limits depending on the initial step.

Even though the outward appearance of some of the governing equations in the isoenergy density theory may be similar to those in ordinary mechanics, the underlying principles and assumptions are fundamentally different.

1.5 Axisymmetric deformation

The deformation of a cylindrical bar specimen stretched uniaxially may be regarded as one of axisymmetry. Only one angle β is needed to locate the position of the isoenergy density element whose corresponding displacements are

$$u_\xi = u_\xi(\xi, \zeta), \quad u_\eta = u_\eta(\eta, \zeta), \quad u_\zeta = u_\zeta(\xi, \zeta). \tag{1.23}$$

The coordinates (ξ, η, ζ) form an orthogonal system and are related to the reference axes (x, y, z). The five nonvanishing deformation gradients are

$$e_\xi = \frac{\partial u_\xi}{\partial \xi}, \quad e_\eta = \frac{\partial u_\eta}{\partial \eta}, \quad e_\zeta = \frac{\partial u_\zeta}{\partial \zeta}, \quad f_\zeta = \frac{\partial u_\xi}{\partial \zeta}, \quad g_\xi = \frac{\partial u_\zeta}{\partial \xi} \tag{1.24}$$

Rate change of volume with surface. It can be deduced from the general expressions in [12] that $(dV/dA)_i$ or $\mathcal{V}_i$ for axisymmetric deformation take the forms

$$\mathcal{V}_\xi = \frac{2e_\eta + 2(e_\xi + e_\zeta)(1 + e_\eta)\cos\beta + 2e_\xi e_\zeta \cos^2\beta - (g_\xi^2 + f_\zeta^2)\sin^2\beta - 2f_\zeta g_\xi}{[2e_\eta(1 + e_\zeta \cos\beta) + 2e_\zeta \cos\beta - f_\zeta^2 \sin^2\beta]\cos\beta}$$

$$\mathcal{V}_\eta = \frac{2e_\eta + 2(e_\xi + e_\zeta)(1 + e_\eta)\cos\beta + 2e_\xi e_\zeta \cos^2\beta - (g_\xi^2 + f_\zeta^2)\sin^2\beta - 2f_\zeta g_\xi}{2[e_\xi + e_\zeta(1 + e_\xi \cos\beta)]\cos\beta - (g_\xi^2 + f_\zeta^2)\sin^2\beta - 2f_\zeta g_\xi} \tag{1.25}$$

$$\mathcal{V}_\zeta = \frac{2e_\eta + 2(e_\xi + e_\zeta)(1 + e_\eta)\cos\beta + 2e_\xi e_\zeta \cos^2\beta - (g_\xi^2 + f_\zeta^2)\sin^2\beta - 2f_\zeta g_\xi}{[2e_\eta(1 + e_\xi \cos\beta) + 2e_\xi \cos\beta - g_\xi^2 \sin^2\beta]\cos\beta}$$

Even though the numerators and denominators of equations (1.25) contain only second-degree terms in the e_ξ, e_η, etc., displacement gradient terms of order higher than second are included in the expansions for $\mathcal{V}_i$ $(i = \xi, \eta, \zeta)$. Because of axisymmetry, the condition

$$\mathscr{V}^\eta = \kappa = \text{const.} \tag{1.26}$$

may be assumed which, when applied to the second of equations (1.25), renders

$$e^\eta = \frac{(\kappa - 1)\left[2\,(e^\xi + e^\zeta)\,\cos\beta + 2e^\xi e^\zeta\,\cos^2\beta - 2f^\zeta g^\xi - (f^2_\zeta + g^2_\xi)\,\sin^2\beta\right]}{2\left[1 + (e^\xi + e^\zeta)\,\cos\beta\right]} \tag{1.27}$$

The condition of isoenergy in equation (1.13) may be further invoked to give

$$2(1 + e^\eta)\,(e^\xi - e^\zeta)\,\cos\beta - (g^2_\xi - f^2_\zeta)\,\sin^2\beta = 0. \tag{1.28}$$

Equation (1.28) solves for β or the position of the isoenergy density element, since e^ξ, e^ζ, f^ζ and g^ξ are known functions of β, i.e.,

$$e^\xi = e_x\,\cos^2\beta + e_z\,\sin^2\beta + (f_z + g_x)\,\sin\beta\,\cos\beta,$$

$$e^\zeta = e_x\,\sin^2\beta + e_z\,\cos^2\beta - (f_z + g_x)\,\sin\beta\,\cos\beta,$$

$$\tag{1.29}$$

$$f^\zeta = (e_z - e_x)\,\sin\beta\,\cos\beta + f_z\,\cos^2\beta - g_x\,\sin^2\beta,$$

$$g^\xi = (e_z - e_x)\,\sin\beta\,\cos\beta - f_z\,\sin^2\beta + g_x\,\cos^2\beta.$$

The quantities e_x, e_z, f_z and g_x are found numerically from the procedure discussed earlier in connection with Figure 1.7.

1.6 Nonequilibrium response of cylindrical bar specimen in tension

Consider a cylindrical bar specimen 10 cm long whose center portion has a diameter of 1.25 cm and its ends are enlarged to 1.45 cm as shown in Figure 1.8. It is made of 6061-T6 aluminum and is displaced at a rate of $\dot{u}_0 = 1.27 \times 10^{-6}$ m/sec. Results are obtained every 120 sec for a total of six steps. A data bank for the 6061-T6 aluminum is constructed that consists of the τ versus e relations for different strain rates covering the behavior of all local elements as illustrated in Figure 1.7. The response of a particular element is determined step by step according to the load history as discussed earlier.

Isoenergy density plane. A reference curve is selected requiring only the initial slope, say $(d\sigma/d\varepsilon)_0 = 8.167 \times 10^5$ Pa for the 6061-T6 aluminum. For a sufficiently small load step, all elements are assumed to be isothermal and possess the same initial stress/strain state. Initial values of $\mathscr{V}_0 = 1$ m and $\beta = 0$ may be used. Upon completion of the first loading step, all elements would

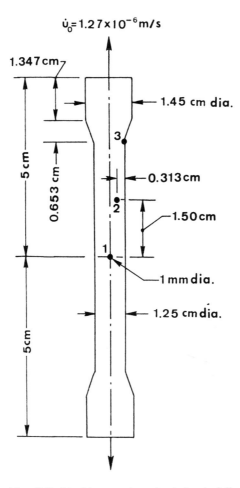

Figure 1.8. Cylindrical bar specimen loaded uniaxially.

experience different stress/strain response and the process would repeat as in the case of a problem with nonuniform initial thermal and mechanical strains. The parameter κ in equation (1.26) is constant only within a given element; it is determined by an iteration procedure such that equation (1.28) is satisfied for each element. The detailed steps can be found in the PEDDA program [16].

In order not to be overwhelmed with numerical data, attention will be focused on the three locations 1, 2 and 3 in Figure 1.8 where the results are averaged over a circular spot of 1 mm in diameter. Table 1.2 summarizes the positions of the isoenergy density elements at point 1, 2 and 3 as time varies from 120 sec to 720 sec. Only small variations of β occur at points 1 and 2. The abrupt change in geometry at point 3 caused a relatively large alteration in β.

Table 1.2. Time history of angle β in radians at points 1, 2 and 3.

Time	Location		
t (sec)	1	2	3
120	0.7854	0.7835	0.7665
240	0.7850	0.7523	0.7604
360	0.7850	0.7636	0.7333
480	0.7852	0.7682	0.7280
600	0.7852	0.7692	0.7272
720	0.7852	0.7689	0.7271

Volume to surface area change. The incremental change of volume with surface area on the isoenergy density plane denoted by $\mathscr{V}$ also alters from element to element. This is shown by the data in Table 1.3. Aside from $\mathscr{V}$ at 1 which remains nearly constant, the change of volume with surface area at 2 and 3 oscillated as time is increased. Their variations in time tend to die out as the equilibrium state is approached.

Table 1.3. Time history of volume to surface area change on the isoenergy density plane at locations 1, 2 and 3.

Time	Location		
t (sec)	1	2	3
120	1.9150	1.9148	1.8967
240	1.9149	1.8589	1.8863
360	1.9155	1.8793	1.8411
480	1.9163	1.8883	1.8338
600	1.9168	1.8870	1.8333
720	1.9172	1.8905	1.8340

Isostress, isostrain and isoenergy. Figure 1.9 displays a plot of τ against e for points 1, 2 and 3. All of the three curves rise sharply and then level off. There are noticeable differences for elements at the different locations. Points 1 and 2 being closer to the center of the specimen, as shown in Figure 1.8, have higher mechanical constraint than point 3 located on the surface. This is why curves 1 and 2 are larger in amplitude than curve 3. The largest τ for a given e occurred at 2. Displayed in Figure 1.10 are the variations of the global isostress with isostrain obtained by averaging all the elements. Refer also to the corresponding numerical values in Table 1.4 in which the volume energy density $\mathscr{W}$ is found by application of equation (1.6).

A salient feature of the isoenergy density theory is the unique representation of the volume energy density function from any one of the nine pairs of (τ_{ij}, e_{ij}) as stated by equation (1.22). For the axisymmetric problem at hand,

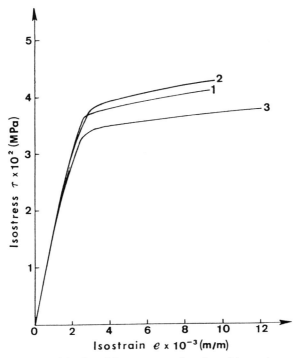

Figure 1.9.   Local isostress as a function of isostrain.

Table 1.4.   Isostress, isostrain and isoenergy density at locations 1, 2 and 3.

Location	Time t (sec)	Isostress τ (MPa)	Isostrain $e \times 10^{-3}$ (m/m)	Isoenergy density W (MJ/m^3)
Point 1	120	256.12	1.6376	0.20971
	240	374.16	2.9759	0.68579
	360	384.43	4.7045	1.3414
	480	395.10	6.4864	2.0360
	600	404.13	7.9953	2.6389
	720	412.41	9.3785	3.2037
Point 2	120	256.30	1.6413	0.21034
	240	379.86	3.1499	0.75718
	360	396.85	4.9638	1.4617
	480	410.16	6.8124	2.2076
	600	418.28	8.3598	2.8485
	720	427.43	9.7784	3.4484
Point 3	120	267.60	1.7530	0.23457
	240	357.59	3.3780	0.79853
	360	355.81	5.8338	1.6724
	480	365.37	8.2496	2.5395
	600	374.83	10.178	3.2498
	720	383.93	11.947	3.9182

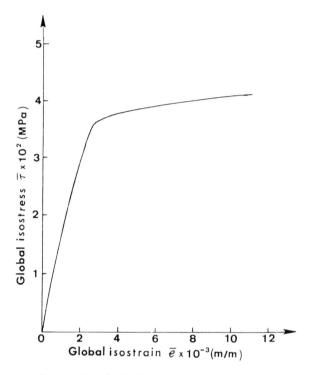

Figure 1.10. Global isostress versus isostrain.

there prevails on the isoenergy density plane five nonzero displacement gradients (e^ξ, e^η, e^ζ, f^ζ, g^ξ) from which the five independent isostress components ($\tau_{\xi\xi}, \tau_{\eta\eta}, \tau_{\zeta\zeta}, \tau_{\xi\zeta}, \tau_{\zeta\xi}$) may be obtained. Since $\mathcal{W}$ and e_{ij} or $e^\xi_\xi, e^{\eta\eta}, ...,$ e^ξ_ζ are already known, τ_{ij} follows immediately from

$$\tau_{ij} = \frac{\partial \mathcal{W}}{\partial e_{ij}}, \quad i, j = \xi, \eta, \zeta. \tag{1.30}$$

Note that nontrivial isostrains are

$$e^{\xi\xi} = \frac{\partial u^\xi}{\partial \xi}, \quad e^{\eta\eta} = \frac{\partial u^\eta}{\partial \eta}, \quad e^{\zeta\zeta} = \frac{\partial u^\zeta}{\partial \zeta}, \quad e^{\xi\zeta} = \frac{\partial u^\xi}{\partial \zeta}, \quad e^{\zeta\xi} = \frac{\partial u^\zeta}{\partial \xi} \tag{1.31}$$

which, according to equations (1.24), are equivalent to

$$e^\xi = e^\xi_\xi, \quad e^\eta = e^{\eta\eta}, \quad e^\zeta = e^{\zeta\zeta}, \quad f^\zeta = e^{\xi\zeta}, \quad g^\xi = e^\zeta_\xi. \tag{1.32}$$

Isostrains are defined in terms of displacement gradients. Hence, once u_i is known, the e_{ij}'s follow from equations (1.31) without loss in generality. That is, the effect of large and finite deformation and are all included*.

Strain transformation. If $\bar{e}_{ij}$ denotes the isostrains referred to the (ξ, η, ζ) axes and e_{ij} to the (r, θ, z) axes, then they are related through the transformations

$$e_{ij} = l_{ki} \, l_{mj} \, \bar{e}_{km}, \tag{1.33}$$

where l_{ij} are the direction cosines. Applying equations (1.33) to the axisymmetric case, the following expressions are found:

$$e_r = e_\xi \, \cos^2 \beta + e^\zeta \, \sin^2 \beta - (f_\zeta + g_\xi) \, \sin \beta \, \cos \beta,$$

$$e_z = e_\xi \, \sin^2 \beta + e^\zeta \, \cos^2 \beta + (f_\zeta + g_\xi) \, \sin \beta \, \cos \beta,$$

$$f_z = -(e_\zeta - e_\xi) \, \sin \beta \, \cos \beta + f_\zeta \, \cos^2 \beta - g_\xi \, \sin^2 \beta, \tag{1.34}$$

$$g_r = -(e_\zeta - e_\xi) \, \sin \beta \, \cos \beta - f_\zeta \, \sin^2 \beta + g_\xi \, \cos^2 \beta.$$

These expressions are, in fact, the inverse of equations (1.29) if e_x, e_z, f_z and g_x are identified with e_r, e_z, f_z and g_r, respectively. They would correspond to the ordinary strains

$$\varepsilon_r = e_r, \quad \varepsilon_z = e_z, \quad \varepsilon_{rz} = \frac{1}{2}(f_z + g_r). \tag{1.35}$$

Their values at points 1, 2 and 3 are given in Table 1.5. As expected, the radial strain is compressive because of necking. The largest strain occurs in the axial direction where the load is applied. Significant shear strains prevail on the bar surface at 3 where the material experiences large distortion.

Calculated in Table 1.6 are the time history of the axial strain rate $\dot{\varepsilon}_z$ at points 1, 2 and 3 and $\bar{\dot{\varepsilon}}_z$ defined as the average of all elements:

$$\bar{\dot{\varepsilon}}_z = \frac{1}{N} \sum_{j=1}^{N} (\dot{\varepsilon}_z)_j, \tag{1.36}$$

---

* It is not necessary to use the definition

$$\varepsilon_{ij} = \frac{1}{2}\left[\frac{\partial u_i}{\partial \xi_j} + \frac{\partial u_j}{\partial \xi_i} - \frac{\partial u_k}{\partial \xi_i} \frac{\partial u_k}{\partial \xi_j}\right], \quad i = 1, 2, 3,$$

as in finite elasticity because the displacement components u_i in the isoenergy density theory are determined independent of the definition of strain.

Table 1.5. Numerical values of ordinary strains at points 1, 2 and 3.

Time t (sec)	ε_r $\times\ 10^{-3}$ (m/m)	ε_z $\times\ 10^{-3}$ (m/m)	ε_{rz} $\times\ 10^{-3}$ (m/m)
Point 1	0.1102 x 10^{10}	3.2698	0
	−0.1568	6.0899	0
	−0.6817	10.040	0
	−1.2296	14.101	0
	−1.6965	17.529	0
	−2.1258	20.663	0
Point 2	0.00631	3.2709	0.00613
	−0.2225	6.5009	0.2226
	−0.7267	10.597	0.2475
	−1.2580	14.771	0.2759
	−1.7220	18.270	0.3646
	−2.1495	21.468	0.3898
Point 3	0.01234	3.4876	0.0657
	−0.2120	6.9436	0.1790
	−0.8025	12.392	0.6899
	−1.4102	17.747	1.1044
	−1.9165	22.021	1.3994
	−2.3763	25.920	1.6571

Table 1.6. Time history of local and global axial strain rate.

Time t (sec)	Local strain rate $\dot\varepsilon_z \times 10^{-5}$ (sec^{-1})			Global strain rate $\dot{\bar\varepsilon}_z \times 10^{-5}$ (sec^{-1})
	1	2	3	
120	2.538	2.709	2.893	2.743
240	2.821	3.053	3.710	3.188
360	3.338	3.446	4.501	3.701
480	3.120	3.197	4.012	3.400
600	2.734	2.790	3.405	2.933
720	2.489	2.540	3.093	2.672

where N denotes the total number of elements used to obtain $\dot{\bar\varepsilon}$. The axial strain rate increased with time reaching a maximum at $t = 360$ sec and then decreases. This trend applies to all locations. The highest $\dot\varepsilon_z$ occurred at point 3 as it is expected.

Stress transformation. The isostress τ_{ij} on the isoenergy density plane shall be distinguished from the ordinary* stress components σ_{ij} as indicated by

---

* The term ordinary is used to identify τ_{ij} with σ_{ij} when $\mathcal{V}_i \to 0$ in the limit although, in general, σ_{ij} no longer transform as components of a second-order tensor. A scalar quantity containing the rate change of volume with surface and body forces is present as shown in equation (1.38).

equation (1.17). The five components of τ_{ij} for the axisymmetric deformation with zero body forces ($h_i = 0$) are given by

$$\tau_{\xi\xi} = \sigma_\xi{}^\xi + \rho\ddot{u}^\xi \mathscr{V}_\xi,$$

$$\tau_{\eta\eta} = \sigma_\eta{}^\eta + \rho\ddot{u}_\eta \mathscr{V}_\eta,$$

$$\tau_{\zeta\zeta} = \sigma_{\zeta\zeta} + \rho\ddot{u}_\zeta \mathscr{V}_\zeta, \qquad (1.37)$$

$$\tau_\xi{}^\zeta = \sigma_\xi{}^\zeta + \rho\ddot{u}^\xi \mathscr{V}_\xi,$$

$$\tau_\zeta{}^\xi = \sigma_\zeta{}^\xi + \rho\ddot{u}_\xi \mathscr{V}_\zeta.$$

Let $\bar{\sigma}_{km}$ refer to the (ξ, η, ζ) coordinate system and σ_{ij} to the (r, Θ, z) system, the transformation law derived in [12] is of the form

$$\sigma_{ij} = l_{ki} l_{mj} \bar{\sigma}_{km} - \rho\ddot{u}_k \mathscr{V}_k \qquad (1.38)$$

Making use of equation (1.38), the ordinary stresses σ_r, σ_z and σ_{rz} are obtained and their values are shown in Table 1.7. From the data in Tables 1.6 and 1.7,

Table 1.7. Variations of ordinary stresses with time at points 1, 2 and 3.

Time t (sec)	σ_r (MPa)	σ_z (MPa)	σ_{rz} (MPa)
Point 1			
120	37.207	475.03	0
240	72.247	676.07	0
360	89.785	679.07	0
480	93.664	696.54	0
600	95.769	712.49	0
720	96.911	727.91	0
Point 2			
120	32.261	480.33	0.8416
240	72.363	687.35	20.360
360	88.194	705.50	13.489
480	94.547	725.77	10.866
600	96.843	739.71	11.723
720	98.629	756.23	10.853
Point 3			
120	45.599	489.60	8.395
240	66.305	648.87	14.575
360	73.241	638.38	29.550
480	80.566	650.17	32.839
600	83.152	666.51	34.104
720	83.241	684.62	35.219

plots of σ_r *versus* $-\varepsilon_r$ are made and presented in Figure 1.11. The trends are similar to those for the isostress and isostrain in Figure 1.9, although the magnitudes are quite different. The same applies to the relationships between σ_z and ε_z. Deviations of the axial stress/strain curves at 1, 2 and 3 from the reference curve are appreciable, Figure 1.12. The effect of inhomogeneity is not

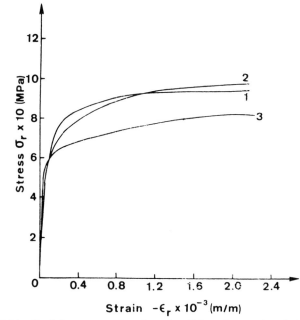

Figure 1.11. Radial stress as a function of radial strain at points 1, 2 and 3.

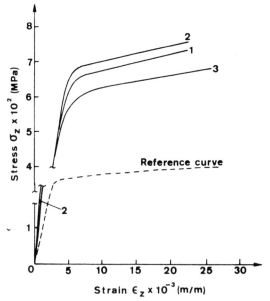

Figure 1.12. Axial stress as a function of axial strain and points 1, 2 and 3.

negligible. An interesting feature of the nonmonotonic variations of σ_{rz} with ε_{rz} occurred at point 2 in Figure 1.13. Such an effect is not intuitively obvious and could be verified by experiments.

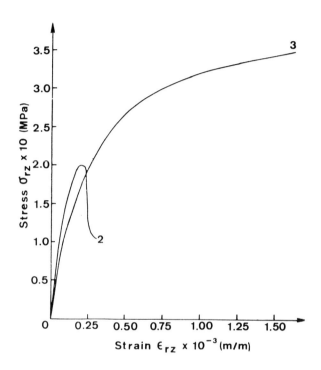

Figure 1.13. Shear stress as a function of shear strain at points 1, 2 and 3.

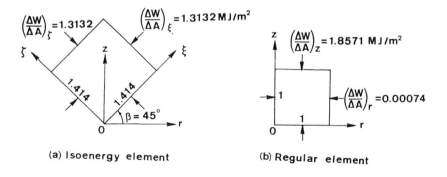

(a) Isoenergy element   (b) Regular element

Figure 1.14. Transmission of surface energy density for isoenergy and regular element at point 1 for $t = 240$ sec.

Surface energy density. The difference between an isoenergy element in contrast to an ordinary element can be best illustrated by the corresponding surface energy densities. Figures 1.14(a) and 1.14(b) referred to point 1 or the center of the cylindrical bar in Figure 1.8. While the same energy is transmitted across all the surfaces for an isoenergy element in Figure 1.14(a) with $(\Delta W/\Delta A)_\xi = (\Delta W/\Delta A)_\zeta = 1.3132$ MJ/m^2 at $t = 240$ sec. The surface energy density $(\Delta W/\Delta A)_z = 1.8571$ MJ/m^2 differed appreciably from $(\Delta W/\Delta A)_r = 7.4 \times 10^{-4}$ MJ/m^2 for an ordinary element. The size of the regular element does not change with location while that of the isoenergy element would be different in addition to its orientation. The angle $\beta = 45°$ at point 1 is the result of geometric and load symmetry at the specimen center. At point 2 in Figure 1.15(a) for the same time instance of $t = 240$ sec, $\beta = 43.1°$ and the normalized

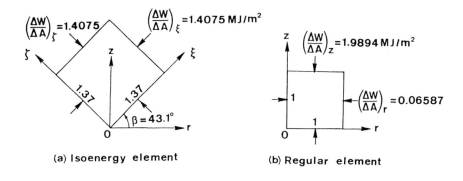

(a) Isoenergy element   (b) Regular element

Figure 1.15.   Transmission of surface energy density for isoenergy and regular element at point 2 for $t = 240$ sec.

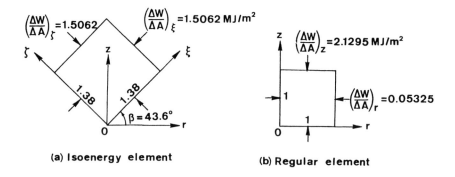

(a) Isoenergy element   (b) Regular element

Figure 1.16.   Transmission of surface energy density for isoenergy and regular element at point 3 for $t = 240$ sec.

isoenergy element size is 1.37 not as large as that at point 1. The difference between the surface energy density in the r and z direction is still large. A similar situation prevails at point 3; it is shown by the results in Figures 1.16(a) and 1.16(b). As time is increased to $t = 720$ sec, Figure 1.17(a) reveals that the isoenergy element size and its orientation at point 1 did not change but the quantity $(\Delta W/\Delta A)^\xi$ or $(\Delta W/\Delta A)_\zeta$ increased considerably. Large difference between $(\Delta W/\Delta A)_r$ and $(\Delta W/\Delta A)_z$ in Figure 1.17(b) is manifestation of uniaxiality. Dependency of the isoenergy element size and its orientation on time can be seen from the the results in Figures 1.18 and 1.19. In retrospect, it

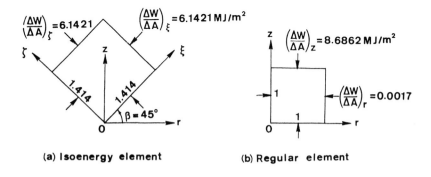

(a) Isoenergy element   (b) Regular element

Figure 1.17. Transmission of surface energy density for isoenergy and regular element at point 1 for $t = 720$ sec.

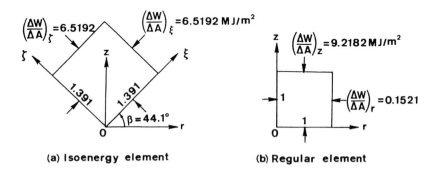

(a) Isoenergy element   (b) Regular element

Figure 1.18. Transmission of surface energy density for isoenergy and regular element at point 2 for $t = 720$ sec.

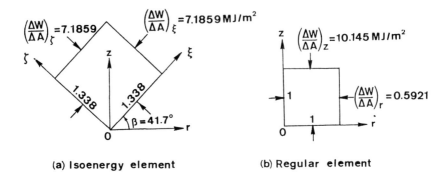

Figure 1.19. Transmission of surface energy density for isoenergy and regular element at point 3 for $t = 720$ sec.

can be said that nonequilibrium leads to inhomogeneity. The values of $(\Delta W/\Delta A)_\xi$, $(\Delta W/\Delta A)_r$ and $(\Delta W/\Delta A)_z$ at other instances of time can be found in Table 1.8.

Table 1.8. Time variations of surface energy density on the isoenergy density and physical plane for points 1, 2 and 3.

Time	Surface energy density (MJ/m^2)		
t (sec)	$(\Delta W/\Delta A)^\xi = (\Delta W/\Delta A)^\zeta$	$(\Delta W/\Delta A)_r$	$(\Delta W/\Delta A)^z$
Point 1			
120	0.4016	1.043×10^{-6}	0.5679
240	1.3132	0.00074	1.8571
360	2.5694	0.00145	3.6337
480	3.9015	0.00109	5.5176
600	5.0582	0.00142	7.1534
720	6.1421	0.00172	8.6862
Point 2			
120	0.4028	0.00108	0.5696
240	1.4075	0.06587	1.9894
360	2.7469	0.08467	3.8838
480	4.1686	0.10138	5.8944
600	5.3751	0.12312	7.6005
720	6.5192	0.15209	9.2182
Point 3			
120	0.4449	0.01189	0.6291
240	1.5062	0.05325	2.1295
360	3.0790	0.22675	4.3485
480	4.6569	0.37781	6.5750
600	5.9578	0.49008	8.4114
720	7.1859	0.59212	10.145

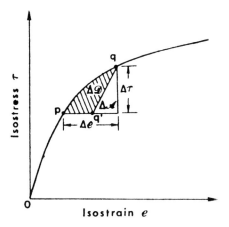

Figure 1.20. Incremental change of dissipation energy density.

Dissipation energy density. Loading and unloading applied to the global stress and global strain of a uniaxial specimen are not the same as the irrecoverability of a local element in a continuum. Tractions and displacements on an element would not vanish when the remotely applied loads are removed. The computation of $\mathcal{D}$ or $\Delta\mathcal{D}$ in equation (1.15) involves the determination of the path qq' in Figure 1.20 that is characteristic of the local recovery property of the material. If the surface and volume energy density are kept within the limits defined in equations (1.1) and (1.2) such that material continuity is preserved locally, the slope of the line qq' could be taken with sufficient accuracy to be proportional to that of the reference state in relation to their corresponding ratios of volume change with surface area, i.e.,

$$\frac{(\Delta\tau/\Delta e)_q}{\mathcal{V}_q} = \frac{(\Delta\tau/\Delta e)_0}{\mathcal{V}_0}. \tag{1.39}$$

Equation (1.39) can be written in general for the kth line of recovery at a given load increment

$$\frac{(\Delta\tau/\Delta e)_k}{\mathcal{V}_k} \rightarrow \frac{(\Delta\tau/\Delta e)_0}{\mathcal{V}_0}, \quad k = 1, 2, \text{ etc.} \tag{1.40}$$

Local recovery is an intrinsic behavior of the material basic constituents* that

---

* When phase transformation [13] occurs $(\Delta\tau/\Delta e)_0$ and $\mathcal{V}_0$ would change accordingly, say from those for a solid to liquid or gas.

is assumed to be retained, even when damage takes place by the creation of new surface and/or fragmentation over a unit volume of material.

Application of equation (1.40) gives $\mathscr{D}$ that increases monotonically with time. This feature is observed in the data shown graphically by Figure 1.21. Energy dissipated is greater at point 3 where distortion is the largest. The lowest corresponds to the specimen center at point 1, whereas the curve for point 2 strikes an intermediate position for it is located in between the specimen surface and its center. Taking all elements into account, the variations of the global average $\bar{\mathscr{D}}$ with time are plotted in Figure 1.22. Dissipation energy is relatively low when the initial temperature of the specimen is below ambient; it increases more rapidly as more heat is generated by mechanical deformation. Refer to Table 1.9 for the numerical values of $\mathscr{D}$ at points 1, 2 and 3.

Temperature change. Once $\mathscr{D}$, $\mathscr{V}$ and e in equation (1.14) are known, Θ follows immediately. The numerical values of $\Theta-\Theta_0$ in Table 1.9 are plotted as a function of time in Figure 1.23. Up to $t = 720$ sec, Θ at points 1 and 3 is still below ambient. This is attributed to the slow rate of loading. The time at which $\Theta-\Theta_0$ switches sign decreases with increasing loading

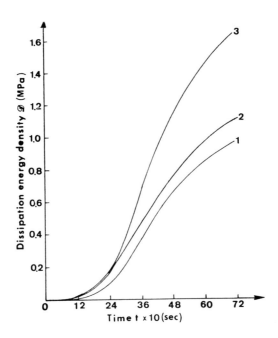

Figure 1.21. Local variations of dissipated energy density with time.

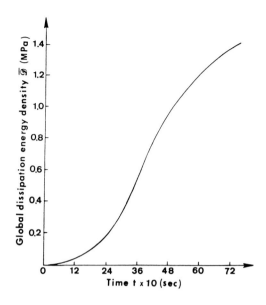

Figure 1.22.   Global dissipated energy density as a function of time.

Table 1.9.   Time variations of dissipation energy density and temperature at 1, 2 and 3.

Time t (sec)	Dissipation $\mathcal{D}$ (MJ/m^3)	Temperature $\Theta-\Theta_0$ ($^\circ$K)
Point 1		
120	0.0031884	−0.13316
240	0.10798	−0.13318
360	0.37370	−0.10144
480	0.66000	−0.08244
600	0.84410	−0.06070
720	0.97880	−0.03443
Point 2		
120	0.042307	−0.06791
240	0.18214	−0.06029
360	0.46546	−0.01426
480	0.77496	0.01075
600	0.96716	0.03381
720	1.1098	0.06641
Point 3		
120	0.027571	−0.07028
240	0.20206	−0.07226
360	0.68863	−0.06126
480	1.1527	−0.04705
600	1.4447	−0.02818
720	1.6859	−0.00711

rate. The global cooling/heating behavior of the 6061-T6 aluminum bar is given in Figure 1.24 which is typical of all slender solids subjected to uniaxial tension.

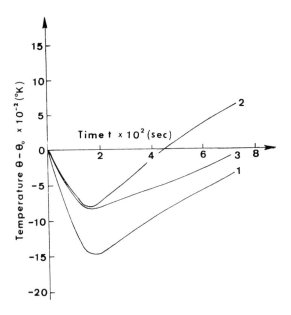

Figure 1.23. Time history of local temperature.

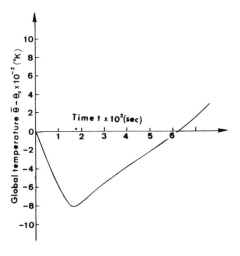

Figure 1.24. Variations of global temperature with time.

1.7 Conclusions

Care should be exercised in distinguishing and/or defining physical processes with reference to their equilibrium or nonequilibrium states. The former infers to systems in which all parts possess the *same* physical properties, both locally and globally. Such a condition is not satisfied in the case of a uniaxial specimen subjected to tension because the interior and surrounding is disturbed continuously. More specifically, the local stress, strain temperature and energy density vary from one part of a system to another at each time instance. The transient character of the constituents cannot be described by referring to the system as a whole unless reference is made to some averages, statistically or otherwise. Such averages on space and time are not always clearly stated in the correlation of experimental data, especially when their interpretation relies on theories that are valid only when the system is in equilibrium. Classical continuum mechanics theories are particularly vulnerable when applied to explain experimental data for *open* systems whose states are, by nature, nonequilibrium.

Application of the isoenergy density theory accounted for the combined effect of size, time and temperature such that the resulting thermal/mechanical behavior can be observed at the macroscopic, microscopic and atomic level without ambiguity. This is because scaling of size/time/temperature is accomplished simultaneously. The time scale would be reduced in accordance with the element size and temperature fluctuation. More details on this can be found in [7]. More precise bounds on the loading of a solid should be established for determining the conditions when the morphology of the continuum would require alteration. This would provide useful information on selecting the time increments for monitoring the thresholds of surface and volume energy density.

References

[1] Sih, G.C. and Tzou, D.Y., "Irreversibility and Damage of SAFC-40R Steel Specimen in Uniaxial Tension", *Theoret. Appl. Fracture Mech.* **7**: (1), 23–30 (1987).

[2] Sih, G.C., Lieu, F.L., and Chao, C.K., "Thermal/Mechanical Damage of 6061-T6 Aluminum Tensile Specimen", *J. Theoret. Appl. Fracture Mech.* **7**: (2), 67–78 (1987).

[3] Nadai, A., "Untersuchungen der Festigkeitslehre mit Hilfe von Temperaturmessungen", Dissertation, Technische Hochschule, Berlin, 1911.

[4] Bottani, C.E. and Caglioti, G., "Mechanical Instabilities of Metals: Temperature Change During Deformation are Indices of Fundamental Processes", *Metallurg. Sci. Technol.* **2**: 1, 3–7 (1984).

[5] Romanov, A.N. and Gadenin, M.M., "Energy Balance for Elastoplastic Fracture: Static and Cyclic Loading", in *Plasticity and Failure Behavior of Solids,* edited by G.C. Sih, A.J. Ishlinsky and S.T. Mileiko, Kluwer Academic Publishers, Dordrecht, 1990.

[6] Cernocky, E.P. and Krempl, E., "A Theory of Thermoviscoplasticity for Uniaxial Mechanical and Thermal Loading", *J. Méc. Appl.* **5**: (3), 293–321 (1981).

[7] Sih, G.C. and Chao, C.K., "Scaling of Size/Time/Temperature – Part 1: Progressive Damage in Uniaxial Tensile Specimen", *J. Theoret. Appl. Fracture Mech.* **12**: (2), 93–108, –.Part 2: Progressive Damage in Uniaxial Compressive Specimen" *J. Theoret. Appl. Fracture Mech.* **12** (2), 109–119 (1989).

[8] Love, A.E.H., *A Treatise on the Mathematical Theory of Elasticity,* 4th edn., Dover Publications, New York, 1944.

[9] Sih, G.C. and Chou, D.M., "Nonequilibrium Thermal/Mechanical Response of 6061-T6 Aluminum Alloy at Elevated Temperature", *J. Theoret. Appl. Fracture Mech.* **12**: (1), 19–31 (1989).

[10] Sih, G.C., "Some Problems in Nonequilibrium Thermomechanics", in *Nonequilibrium Thermodynamics,* edited by S. Sieniutycz and P. Salamon, Taylor and Francis, New York (in press).

[11] Sih, G.C., "Mechanics and Physics of Energy Density and Rate of Change of Volume with Surface", *J. Theoret. Appl. Fracture Mech.* **4** (3), 157–173 (1985).

[12] Sih, G.C., "Thermomechanics of Solids: Nonequilibrium and Irreversibility", *J. Theoret. Appl. Fracture Mech.* **9**: (3), 175–198 (1988).

[13] Sih, G.C., "Thermal/Mechanical Interaction Associated with Micromechanisms of Material Behavior", Institute of Fracture and Solid Mechanics, Lehigh University Technical Report, U.S. Library of Congress No. 87-080715, February 1987.

[14] Lieu, F.L., "Experiments on Cooling/Heating of Aluminum Tubes under Torsion", Institute of Fracture and Solid Mechanics, Lehigh University, 1986.

[15] Cernocky, E.P. and Krempl, E., "A Coupled, Isotropic Theory of Thermoviscoplasticity Based on Total Strain and Overstress and its Prediction in Monotonic Torsional Loading", *J. Thermal Stresses* **4**, 69–82 (1981).

[16] Sih, G.C. and Tzou, D.Y., "Plane Energy Density Damage Analysis", Institute of Fracture and Solid Mechanics Technical Report IFSM-84-131, 1984.

Thoughts on energy density, fracture and thermal emission

2.1 Introduction

Fracture mechanics has developed in an attempt to understand the fracture of materials and to translate this understanding into structural failure prediction. Research workers in this field have searched for the fundamental quantities governing structural behaviour and thermomechanics. As a result, present theories of fracture have been formulated in terms of stress, strain and energy. One approach which is extensively used is based on a critical energy density [1, 2]. Advantages of energy density theory are that it may be applied to both notched and unnotched specimens [1–6], and can be used in both the elastic and the elasto-plastic regime. A detailed review of this theory is presented in [3] whilst several applications are presented in [7–9].

In the linear elastic regime, the coupling between mechanical and thermal energies is now well established [10, 11]. Areas experiencing a high energy field also experience rises, from ambient, of the local temperature field and these temperature changes can be measured with an infra-red camera [12, 13].

This report will illustrate how energy density theory, together with surface temperature measurements, can be used to assess damage, and in one instance, to develop an optimum local redesign. Attention will also focus on how the temperature field, resulting from the mechanical loading, can be used to confirm this assessment as well as to evaluate the effect of this redesign.

2.2 The F-111 wing pivot fitting

The strain energy density criterion is thought to apply to all material regardless of its behaviour [1]. In this approach, the minima and maxima of dW/dV, which is defined as

G. C. Sih and E. E. Gdoutos (eds): Mechanics and Physics of Energy Density, 35–58.
© 1992 *Kluwer Academic Publishers. Printed in the Netherlands.*

$$\frac{dW}{dV} = \int_0^{\varepsilon_{ij}} \sigma_{ij} \, d\varepsilon_{ij} \tag{2.1}$$

are used to locate the failure sites corresponding to excessive dilatation and distortion, respectively, even when gross yielding takes place.

Failure under monotonic loading is thus assumed to initiate at the site where the local minimum of dW/dV is a maximum, i.e., $(dW/dV)_{\min}^{\max}$ and occurs when a critical value, say $(dW/dV)_c$, is reached.

If the load is applied repeatedly, the hysteresis energy density accumulated for each cycle is thought to control failure initiation.

In a recent paper [7], it was shown that energy density theory is able to explain several phenomena associated with fatigue life enhancement techniques. In the problem considered, it was demonstrated that during the loading/unloading process, the increment dW of the local energy density becomes out-of-phase with respect to the increment in the global work. This "change in phase" was a necessary condition for fatigue life enhancement. However, it will be shown that, depending on the subsequent load history, this change does not necessarily result in a fatigue life enhancement.

As an illustration of this philosophy, let us consider the local stress and strain fields in the stiffener runout of the F-111 Wing Pivot Fitting (WPF) during a Cold Proof Load Test (CPLT). As a result of cracks in the integral stiffeners under the upper surface of the WPF, the Aeronautical Research Laboratories (ARL) was tasked to design a suitable reinforcement scheme. Reference [8] describes the design concept for a boron/epoxy doubler bonded to the upper surface of the WPF over the critical area.

As a result of this design study, a detailed two dimensional finite element model of the unreinforced WPF was developed (see [8] for details). A view of the critical section is shown in Figure 2.1.

This paper examines the behaviour of the local energy density in the vicinity of the critical region during the CPLT. An elastic-plastic analysis was conducted in which the model was subjected to incremental loading to a maximum load corresponding to 7.3 g and a minimum load corresponding to -2.4 g, (see [8] for details). At each load increment, the energy density was calculated at the nodal points shown in Figure 2.2. Some typical results are shown in Figure 2.3.

It can be seen that, for each nodal point, on unloading the energy density changes phase with respect to the global load, i.e., as the load is removed a stage is reached after which a further decrease in load gives rise to an increase in the local energy. However, loading from 0 to -2.4 g gives rise to an increase in energy density. The magnitude of this increase becomes significantly larger towards the re-entrant corner (node 267, see Figure 2.2). The energy density at node 267 is such that a small flaw at this location may cause the critical energy density to be exceeded. This explains the failures occurring in the CPLT.

As in the fatigue life enhancement procedure which involved cold working of fastener holes discussed in [7], the phase change in energy density is due to prior plasticity. However, unlike cold working, the effect of this phase change is detrimental. This clearly shows that the phase change phenomenon reported in [7] is a necessary but not a sufficient condition for fatigue life enhancement.

In fracture related problems, the observed phase change of the local energy

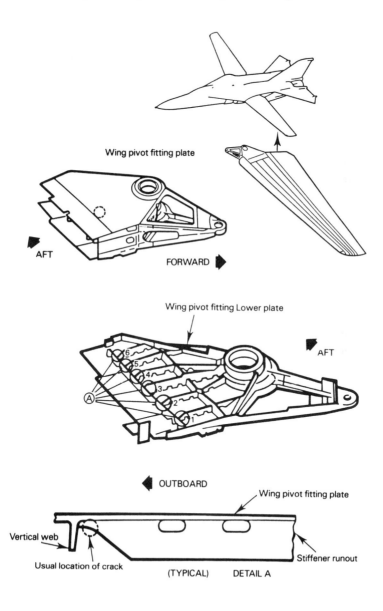

Figure 2.1. F-111 wing pivot fitting, location of critical region of stiffener runout section.

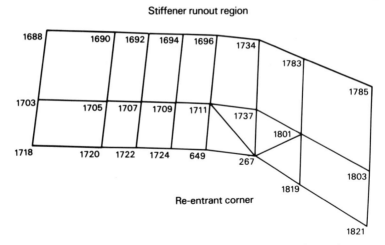

Figure 2.2. Critical region, finite element mesh showing node numbers.

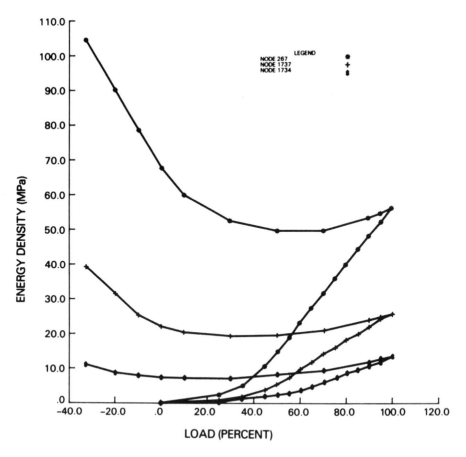

Figure 2.3. Plot of energy density versus load for nodes through the depth of the stiffener.

density with respect to the global work means that current J integral methods and other extensions of linear elastic fracture mechanics, which are based on global behaviour (i.e., load point displacements, etc.), may sometimes be invalid. Only in the cases where the global behaviour is closely related to the local crack tip behaviour can J be considered valid. Energy integrals which are more closely related to the behaviour of the material at the crack tip are discussed by Atluri [14].

Implications for repair. From Figure 2.3, it can be seen that the major contribution to the accumulated energy density field is due to the load sequence, 0 g: 7.3 g: 0 g. In order to lower the residual energy density obtained on unloading, it is necessary to rework the local geometry of the stiffener runout. Repairing, by means of the bonded doubler outlined in [8], without reworking of the stiffener runout will merely lower the energy density that is added to the already very high residual energy. The optimum solution is to both rework the stiffener runout and reinforce the wing pivot fitting. A preload applied to the wing may also be used to lower the residual energy density. It is recommended that the stiffener runout be reworked in conjunction with doubler application; see [8] for details.

2.3 Damage assessment of an F/A-18 stabilator

Background. As a second example, consider the repair of a damaged F/A-18 horizontal stabilator which has recently been incorporated into the Composite Repair Engineering Development Program (CREDP). CREDP is a joint program between the Canadian Forces (CF), the RAAF and the United States Navy (USN) to evaluate the repair capability of damaged composite components on the F/A-18. As part of this program, the Aeronautical Research Laboratory (ARL) was tasked to develop and analyze a repair for the damaged horizontal stabilator, which was initially classed as unserviceable and unrepairable due to two fragment strikes from a tracer rocket.

The F/A-18 horizontal stabilator comprises graphite/epoxy skins with fibres oriented in the 0° and ±45° directions. The skin is supported on a full-depth honeycomb and the stabilator is fully symmetric top and bottom. In the region of the damage, the skin is 29 plies thick (3.68 mm), but tapers off to 7 plies (0.89 mm) at the leading edge.

The horizontal stabilator has incurred extensive local damage to the composite skin and the underlying honeycomb, but neither fragment completely penetrated to the other side of the stabilator. As seen in Figure 2.4, the two damaged areas are approximately the same size. However, there was a slight difference in the impact angle of the two fragment strikes. The location of the damaged zone and ply orientation can be seen in Figure 2.5.

Figure 2.4. Damage zone on the F/A-18 horizontal stabilator.

Structural evaluation. In order to evaluate the structural significance of the damage, a test rig capable of applying the loads was required. Fortunately, ARL had previously developed a test rig for a dynamic test on the horizontal stabilator, see Figure 2.6. The stabilator was mounted in the rig by means of the spindle, which is used to connect the stabilator to the aircraft. All bending loads are transferred to the rig via the spindle. A lever arm connected to the root of the stabilator transfers all torque to the test rig.

The static loads applied to the surface were achieved by using two Firestone air bags resting on the surface of the stabilator. An additional structure, consisting of steel I beams, was built around the test rig to allow the air bags to be mounted above and below the stabilator. Prior to this test, a pressure versus extension calibration test was performed on the air bags to determine the force that the air bags applied to the stabilator.

Four strain gauge rosettes were located in a rectangular pattern around the damage zone and in the corresponding location on the undamaged skin. The gauges in each rosette were aligned in the direction of the $0°$ and $\pm45°$ fibres; see Figures 2.5 and 2.7. Two displacement transducers were placed half way between the left-hand inboard rosettes (1 and 2), in the $0°$ fibre direction, as seen in Figure 2.7. The gauge length of the displacement transducers was 385 mm. The strain gauge rosettes and displacement transducers were positioned so as to evaluate the change in compliance of the structure due to the major damage.

Displacement and pressure transducers were required to measure the extension and pressure for each air bag. The force applied by each air bag is a function of these two variables and was read off a calibration graph. The root bending moment (RBM) was calculated and used as the reference load applied to the structure. Three extra displacement transducers were attached to the

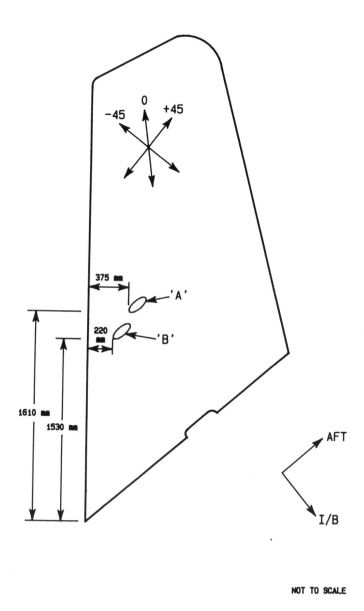

Figure 2.5. Damaged F/A-18 horizontal stabilator.

Figure 2.6.   F/A-18 horizontal stabilator test rig with air bags mounted above.

stabilator at the tip, leading and trailing edges, to measure tip displacement and torque induced by the applied load. A HP9816 data acquisition system was used to record the resulting 33 channels of data.

Test description. The first stage of the investigation involved the application of a static load by incrementing the air bag pressure in steps of 10 kPa. The maximum load applied to the structure was not considered to be critical as the primary objective was a comparison between the strains and the compliance on the top and bottom surfaces of the stabilator. Initially, the air bags were placed underneath the stabilator, producing tension in the damaged surface skin, and a dummy loading and unloading run was performed to allow the structure to settle. Several loading and unloading runs were made in this configuration with strain gauge and transducer readings being conducted at each increment of pressure. The loading was then repeated with the air bags re-configured above the stabilator producing a compressive load in the damaged surface.

Results and discussion. The Ultimate Bending Moment (UBM) and the Design Limit Bending Moment (DLBM) for the horizontal stabilator are 1065 kip-in (120.3 kNm) and 710 kip.in (80.2 kNm), respectively, [28]. The RBM was calculated for each load increment and the test achieved 31% (47%) and 26% (39%) UBM (DLBM) in tension and compression, respectively.

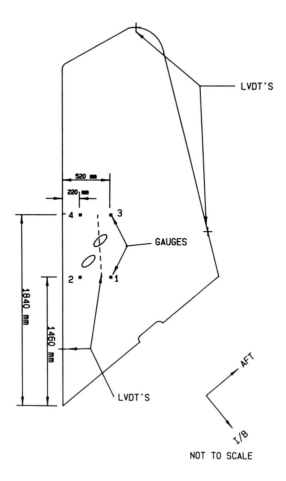

Figure 2.7. Location of strain gauges and displacement transducers.

   The strain survey results for one loading cycle are shown in Table 2.1 for the case of a compressive load only, the results for tensile loading are very similar. All strain values are presented in microstrain. Table 2.1 shows that there is, in most cases, no structurally significant difference between the top and bottom surface strain gauge result in the 0° fibre direction. However, there is a 24% to 27% reduction in gauge D2 0°, which is located close to the edge of the damage "A", depending upon the value of the RBM. This may be due to the gauge being shielded by the damage. The compliance readings, shown at the bottom of the tables as "U strain" for the undamaged surface strain and "D strain" for the damaged surface strain, result in no significant difference between top and bottom surfaces. This implies that the local compliance has not been affected by the damage and indicates that there is little structural degradation.

Table 2.1 Strain ($\mu\varepsilon$) results for damaged and undamaged skins: damaged surface in compression.

R.B.M. (kNm)	0	−9	−18	−25	−31	−25	−18	−9	0
Torque (kNm)	0	−6	−13	−18	−22	−18	−13	−6	0
Gauge D1 +45	−1	−131	−258	−361	−453	−369	−269	−144	−1
Gauge D1 0	−1	−263	−516	−717	−899	−732	−531	−283	−1
Gauge D1 −45	0	−49	−95	−132	−164	−132	−96	−51	−1
Gauge D2 +45	−1	−64	−125	−175	−219	−176	−127	−69	−1
Gauge D2 0	0	−119	−232	−321	−400	−327	−240	−129	−1
Gauge D2 −45	−1	−12	−24	−32	−39	−33	−30	−17	−1
Gauge D3 +45	0	−182	−356	−496	−622	−505	−366	−194	−1
Gauge D3 0	−1	−423	−827	−1151	−1437	−1156	−831	−434	−2
Gauge D3 −45	0	−95	−186	−259	−327	−262	−191	−101	−2
Gauge D4 +45	−1	−126	−246	−344	−431	−351	−256	−138	−1
Gauge D4 0	−1	−261	−508	−705	−878	−717	−522	−281	−1
Gauge D4 −45	−1	−25	−47	−63	−78	−64	−48	−27	−1
Gauge U1 +45	0	199	232	322	401	328	238	127	−1
Gauge U1 0	−1	260	507	705	877	714	518	275	−1
Gauge U1 −45	0	60	119	165	204	164	118	60	−1
Gauge U2 +45	−1	113	219	304	376	308	224	119	−1
Gauge U2 0	−1	161	316	440	550	446	320	168	−1
Gauge U2 −45	−1	−16	−27	−36	−42	−39	−37	−24	−1
Gauge U3 +45	−1	171	333	463	573	467	339	180	−1
Gauge U3 0	−1	405	790	1097	1357	1107	804	429	−1
Gauge U3 −45	−1	112	218	302	373	304	220	117	−1
Gauge U4 +45	−1	93	182	252	311	252	181	95	−1
Gauge U4 0	−1	257	502	702	872	708	513	272	0
Gauge U4 −45	0	−3	−6	−8	−10	−8	−6	−3	0
U Strain	1	370	734	1038	1285	1067	788	438	0
D Strain	0	363	−719	−1016	−1260	−1041	−754	−403	0

2.4 The finite element model

The damaged section of the horizontal stabilator was modelled by representing the skin as a two-dimensional membrane, with effective laminate properties, and the underlying honeycomb as three-dimensional isoparametric brick elements. One proposal for the repair to the damaged zone of the horizontal stabilator involves the application of an external doubler. In this case, the adhesive layer was modelled using three-dimensional isoparametric brick elements and the doubler was represented by two-dimensional membrane elements, with an effective laminate property. Due to the symmetry considerations, only a quarter of the structure was modelled and restraints were applied to represent the level of restraints existing in the structure. Three cases were analyzed, the unrepaired and undamaged skin, the damaged but unrepaired skin and the damaged and repaired skin. The damaged unrepaired model contained 520 nodes, 128 elements and 1255 degrees of freedom and the damaged repaired model contained 818 nodes, 256 elements and 1976 degrees of freedom. The plan view of the mesh used for the three analyses can be seen in

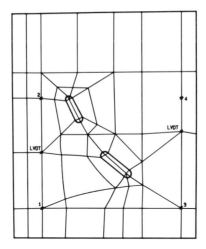

Figure 2.8. Plan view of finite element mesh.

Figure 2.8. From Figure 2.4, it can be seen that a number of surface plies have been delaminated from the skin around the points of penetration. The initial finite element model does not include this effect.

The material used for the skin and the doubler was AS4/3501-6 and the associated laminate properties can be found in Table 2.2. The Young's modulus of the honeycomb was taken to be 2000 MPa with a Poisson's ratio of 0.35. The Young's modulus of the adhesive was taken to be 1800 MPa with a Poisson's ratio of 0.35. The doubler was taken to be 16 plies (2.032 mm thick). A pressure of 400 MPa was applied along one edge of the skin.

Table 2.2. Laminate properties for the horizontal stabilator skin and repair doubler.

	E_{XX} (MPa)	E_{YY} (MPa)	G_{XY} (MPa)	v_{XY}	v_{YX}
Skin	66,708	32,542	14,499	0.3889	0.1897
Repair	57,423	32,863	17,228	0.4428	0.2534

Calibration of the finite element model. The accuracy of the finite element model and the materials properties assumed for the honeycomb structure were evaluated by comparing the predicted with the measured strains, obtained in [15]. For the undamaged stabilator, the ultimate design bending moment [15], corresponds to a design limit strain ε_{dls} of 4850 $\mu\varepsilon$.

The numerical model was configured so that nodes corresponded to the location of the linear voltage displacement transducer (LVDT) and strain gauges positions in the stabilator test [15]. The corresponding values of strain, at the two inboard strain gauges, and the LVDT were close to the measured values, but because the actual structure tapers off towards the leading edge, the strains in the numerical model closest to the leading edge (gauges 2 and 4) were higher than those achieved in the experiment.

Results and discussion. Following the calibration of the undamaged model, the cases were considered in which the model contained a representation of the damaged region. The first case simulated the unrepaired damaged stabilator and the second represented an external doubler applied over the damage region. A summary of the resulting strains over the LVDT and strain gauges, in the 0° fibre direction, is shown in Table 2.3.

Table 2.3. Summary of strain results from F.E. analysis.

Gauge	Unrepaired $\mu\varepsilon$	Repaired $\mu\varepsilon$	Percentage reduction
1	4,953	3,557	28.2
2	4,621	3,390	26.4
3	5,210	3,658	29.8
4	4,863	3,522	27.6
LVDT	5,284	3,664	30.7

The application of an external doubler was found to reduce the strain around the damaged zone. The average reduction in strain was 28.5%. The adhesive shear stresses for this repair were less than 25 MPa, which is below the design allowable of 40 MPa. For the unrepaired case, the stress concentration around the damage was found to be very localized, see Figure 2.9, and is consistent with the experimentally measured thermal emission profile, see [15], and section 2.5 as well as the experimentally measured values of strain and compliance.

ENERGY AROUND DAMAGED AREA

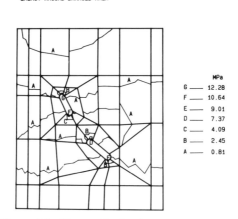

MPa	
G	12.28
F	10.64
E	9.01
D	7.37
C	4.09
B	2.45
A	0.81

Figure 2.9. Energy distribution in stabilator skin.

2.5 Thermoelastic evaluation of damage

The theory describing the coupling between mechanical deformation and thermal energy of an elastic body was first published in 1855 by Lord Kelvin. For an elastic isotropic material undergoing cyclic loading, this theory states that under adiabatic conditions, temperature increments are a linear function of the sum of the principal stresses, and are independent of loading frequency and mean load. Therefore, measurement of the temperature change at a point of a body will define the bulk stress at that point due to applied cyclic loading, if the conditions are adiabatic. Adiabatic conditions will be achieved if the frequency of loading is such that there is negligible heat transfer from one point to another during the period of a cycle. Typically, for metals, this requires a loading frequency of 5 Hz or more.

A recently developed non-contact device marketed under the trade name SPATE 8000 allows temperature measurements to be made for a large number of points on the surface of a cyclically loaded body. These temperature measurements are made by using an infra-red sensitive detector. The SPATE system has the potential for a spatial and temperature resolution of 0.25 mm^2 and 0.001 °K, respectively. Therefore, based on the thermoelastic effect, SPATE output can provide surface stress information for a cyclically loaded structure. This approach has been applied to qualitatively and quantitatively identify and evaluate areas of high stresses or stress gradients in metals for a number of problems [10–12].

Recent experimental work by Machin *et al.* [13] has revealed a second order effect, whereby the temperature response is, in fact, also dependent on mean load. Subsequently, Wong *et al.* [18] have re-derived Kelvin's equation and shown that this effect is due to the slight temperature dependence of the elastic properties of the material. For a mean load of zero, the new formulation reduces to Kelvin's equation. Further experimental results have also validated this revised theory [19]. Thermoelastic theory has now also been generalised for an anisotropic material [20] and some experimental work on composites has been done [21, 22]. Probably due to the recent development of this method, few applications have been found in the literature to bonded repairs (either metals or composites) or to assessing damage in composites [23, 24]. At present, to assess the structural significance of damage, it is usually necessary to extensively strain gauge a specimen and perform tests to reproduce as closely as possible, the local geometry, degree of local constraint and operational stresses of the aircraft component. Clearly, the use of a noncontact approach providing stress information over a large region may have many advantages.

For metals, there has been some work concerning damage assessment in which Chan *et al.* [25] has determined stress intensities of cracked plates. There are some difficulties in the analysis of such problems, due to the spatial resolution of the measurements in relation to the high stress gradient near a crack.

It should be noted that SPATE output at this stage gives only a weighted sum of the principal stresses and not the individual stress components. However, very recently, there has been interest in developing hybrid analytical/experimental approaches to extract individual stress components from SPATE data, as discussed later on.

F-111 stiffener runout replica specimen. As discussed in section 2.3, the Aeronautical Research Laboratory (ARL) has for some time been concerned with the maintenance of the structural integrity of F-111C aircraft in service with the Royal Australian Air Force (RAAF). Particular attention has been focused on the failure of the Wing Pivot Fitting (WPF) under Cold Proof Load Test (CPLT) conditions.

The structure of the wing is such that a 2024–T856 wing skin is spliced onto a D6ac WPF by a series of flush fasteners. Cracking is occurring inboard from this splice due to high stresses induced during certification-testing of the aircraft in CPLT. To overcome this problem, a reinforcement scheme consisting of two discrete boron/epoxy doublers bonded to the upper surface of the WPF over stiffeners known to be subjected to high stresses, has been developed and tested [8]. The doubler reinforcement, when bonded to the upper WPF surface, considerably decreases the stress magnitude in the stiffener runout region. However, as outlined in section 2.3, in order to reduce the stresses and energy in the critical region to acceptable levels, it is also necessary to rework the local geometry of the stiffener runout region.

It was initially proposed to investigate this suggestion utilising the test wing used in the reinforcement substantiation program, but unfortunately, this was rendered unusable. As a result, structurally representative specimens were used. This section presents the results of an investigation of various stiffener runout configurations, utilising a representative geometry stiffener replica specimen. A quantitative measure of the effect of various grind out radii was made using a SPATE 8000 (Stress Pattern Analysis by measurement of Thermal Emission).

Test specimen configuration. The test specimens used in this work were identical to those used in an earlier reinforcement effectiveness study [26]. The specimens were manufactured to represent the area of the WPF prone to cracking (Figures 2.1 and 2.9). As in the wing, steel and aluminum were used where appropriate. Details of the specimen are given in [26].

In the present investigation, the specimens were mounted into hydraulic grips of an Instron 500 testing machine and SPATE scans were obtained for a number of stiffener runout reworks. The rework consisted of grinding out the runout region sequentially to radii of 4, 11 and 22 mm (see Figure 2.10), using conventional workshop grinding wheels. In each case, a smooth transition was achieved between the radius of the stiffener.

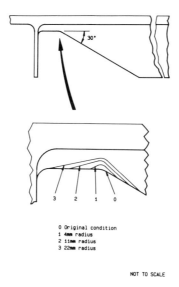

0 Original condition
1 4mm radius
2 11mm radius
3 22mm radius

NOT TO SCALE

Figure 2.10. Stiffener specimen runout rework.

SPATE analysis and results. As previously mentioned, the SPATE detector responds linearly to changes in infra-red flux resulting from small changes in temperature as the specimen is adiabatically cycled. As the flux emitted by a surface is dependent upon its emissivity, which is relatively low (commonly 0.15 for bare metals), the areas to be investigated on the specimen were sprayed with a matt black paint to increase the surface emissivity (typically to 0.92).

The specimen was then placed into the testing machine with the flexural frame attached and the SPATE detector was located within close proximity. A constant amplitude cyclic load of -50 to -100 kN was then applied to the ends of the specimen. A frequency of 6 hertz was chosen to ensure adiabatic conditions. The number of cycles for a given area "scan" was dependent upon the size of the region and the number of grid points chosen.

When viewing the runout region of the unreworked (i.e., initial configuration) specimen, it was shown that the results were reproducible regardless of the viewing direction. The specimen was also removed and subsequently remounted in the testing machine and it was demonstrated that reproducible results were achieved. Following this, the specimen was removed, ground to the new runout configuration and SPATE scans taken. It should be noted that uncalibrated stresses are presented in the figures contained in this work. Furthermore, the contoured data were post-processed using a "smoothing" algorithm, which is described in [27], to eliminate thermal and experimental "noise".

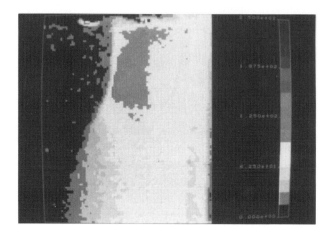

Figure 2.11. Uncalibrated stress analysis of original stiffener geometry with heavy restraint (side view).

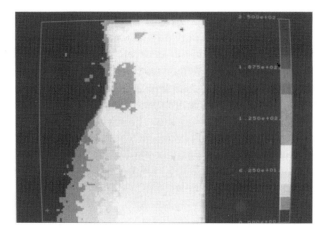

Figure 2.12. Uncalibrated stress analysis of 22 mm runout grind radius heavy restraint (side view).

Results for the unreworked specimen are shown in Figure 2.11 while those for the 22 mm rework radii are given in Figures 2.12. It can be seen from Figures 2.11 and 2.12 that a stress reduction of approximately 15% resulted when a local 22 mm grindout radius at the stiffener runout was compared to the original geometry of a shape re-entrant corner. This confirmed earlier numerical analysis and aided in the choice of the optimum reworked geometry.

F/A-18 stabilator. In general, a region of high energy density, often referred to as a hot spot, will give rise to a region with a large thermal emission. In particular, a hole delamination damaged or a crack in a uniformly stressed component can be immediately seen as it results in a large change in the thermal emission profile. If damage is not structurally significant, i.e., it does not result in a significant change in the local stress field, it will not result in a significant change in the thermal emission profile.

As a result, the thermal emission profile is thought to be a particularly valuable tool in assessing the structural significance of damage, in contrast to other more "passive" non-destructive techniques such as C-scan, X-ray, etc., which only provide information on the geometry of the damage and not its structural significance. Indeed, the thermal emission profile "activity" reflects the interaction of the damage with the structure, material, local geometry and load in a non-destructive fashion. As such, it is particularly well suited to the present investigation and will be used to evaluate the structural significance of the fragment damage and to locate other damage locations.

In order to carry out the structural tests on the horizontal stabilator, the same rig as described earlier was used. To investigate the stress concentration around the damage, thermal emission scans were taken of the damaged region. Two electromagnetic shakers were attached to the underside of the stabilator, at a position that allowed the fundamental bending mode, at approximately 14 Hz, to be excited. A mirror was used to reflect the resultant thermal emission to the SPATE detector unit. The mirror was isolated from the test rig and floor to avoid the problem of vibration.

To further investigate the stress concentration around the damage, a number of detailed thermal emission scans were taken. The results of one of these scans is shown in Figure 2.13, and to assist the reader in interpreting the picture, the damaged region is marked. Examination of this figure reveals that, as indicated by the analysis and by the strain gauge and compliance measurements, there was no "classical" stress concentration affect around the damage. There was a region, in the vicinity of gauge U2, which gave a spurious thermal emission reading. This spurious reading vanished when the structure was excited at 12 Hz.

In conclusion, the thermal emission results taken in conjunction with the results of the finite element analysis and the experimental results presented earlier indicate minimal structural degradation resulting from the fragment strikes, and do not show the classical stress concentration around the fragment holes. This is consistent with measurements of local compliance using strain gauges, discussed in detail in [28]. Therefore, since the damage does not

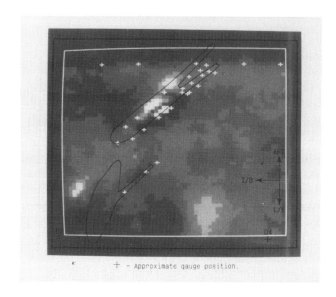

Figure 2.13.   SPATE picture of the damage zone.

appear to compromise the compliance of the structure, a simple externally bonded repair should suffice.

To confirm the numerical and experimental investigations, a fourteen ply external repair was bonded to the stabilator. This repaired stabilator has now been loaded to 120% of design limit load with no apparent structural degradation.

2.6 Stress fields from temperature measurements

In recent years, the thermoelastic effect has been greatly exploited as a non-contact method of experimental stress analysis. The SPATE apparatus, used at ARL, accomplishes this by measuring the thermal emission of the surface of the component which is being sinusoidally loaded about some mean load. The thermal emission is proportional to the surface temperature and, because the thermal fluctuations are proportional to the changes in bulk stress $(\sigma_{xx} + \sigma_{yy} + \sigma_{zz})$, a strain gauge can be used to calibrate the output from SPATE in order to determine the actual bulk stress.

Although, as illustrated by the previous section, the use of SPATE for the analysis of components is a powerful and versatile technique, the vectorial nature of stress is ignored since temperature and bulk stress are scalar values. This places a significant limitation on the ability of SPATE data to provide a quantitative measure of the stress field, except in circumstances where it is known that the stress field is primarily unidirectional. Another disadvantage is that no direct information concerning the nature of the internal stress field of a three-dimensional structure can be obtained from the simple surface analysis performed by SPATE.

However, it is well-known that the stress field existing in a component must satisfy equilibrium and any imposed boundary conditions. For two-dimensional problems, Ryall and Wong [29] have shown that when additional stress-related information is available, such as experimentally obtained bulk stress data from a SPATE analysis, it is no longer necessary to have a complete knowledge of the boundary conditions governing the problem in order to obtain a useful solution. The method involved the use of an assumed stress potential with unknown coefficients, from which the bulk stress could be calculated analytically. The values of the unknown coefficients were determined by formulating the problem using a best-fit criterion where the difference between the analytical expression of the bulk stress and the experimentally observed bulk stresses were minimized in the least squares sense.

This section will extend this formulation to three dimensions, thus enabling the energy field to be calculated. A knowledge of the energy field can be used to locate fracture critical points or used to redesign, so as to achieve an "optimum design".

In order to evaluate the stresses in a body from a knowledge of the surface temperature field, it is first necessary to formulate an energy functional which treats the bulk stress as an independent degree of freedom.

The energy functional. One such approach which uses a three-dimensional penalty element, is given in [30] and uses the bulk stress at each node as a degree of freedom, in addition to the usual u, v and w displacements. The penalty functional method, as described in [30] utilizes a modified energy π functional that is constrained by the equation for bulk stress

$$\pi = 1/2 \int \sigma_{ij} \varepsilon_{ij} dv + 1/E \int (\sigma_\beta - K \varepsilon_\beta)^2 dv. \tag{2.2}$$

Here, σ_β, ε_β and K are the bulk stresses dilatation and bulk modulus, respectively, whilst E is the Young's modulus.

Validation of penalty element formulation. Consider the problem of an edge crack in a rectangular specimen 20 mm long, 10 mm wide and 2 mm thick. The crack is 1.666 mm long and is shown in Figure 2.14. The problem is assumed to behave elastically and Young's modulus and Poisson's ratio of the material are 1.89 GPa and 0.35, respectively. The finite element model consisted of approximately 50, three-dimensional penalty elements as shown in Figure 2.15.

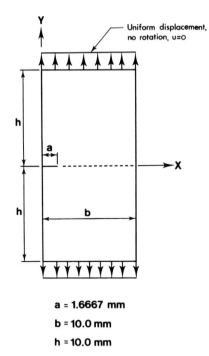

a = 1.6667 mm

b = 10.0 mm

h = 10.0 mm

Figure 2.14. Edge-cracked panel showing main dimensions.

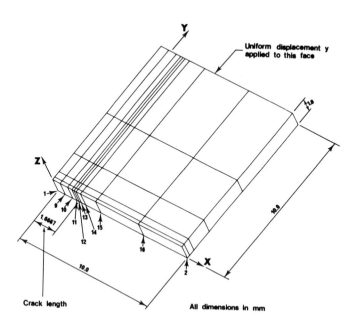

Figure 2.15. Finite element mesh used to model the edge-cracked panel.

Each element had 20 nodal points and used a quadratic form for the displacement field and the bulk stress. This approach gave a value of K_1 of 0.142 MPa $\sqrt{m}$ at the free surface and 0.145 MPa $\sqrt{m}$ at the midsurface which compares very well with the analytical solution of 0.145 MPa $\sqrt{m}$.

As a second problem, the values of the bulk stress on the external surfaces as previously calculated were used as input data together with the surface strains at three equally spaced points on each external surface. For this problem, neither the loads or the displacements were specified. The formulation described above was able to determine the individual stress components, the energy density field, the displacement field and the total load applied to the structure.

The accuracy of the solution can be judged from Figures 2.16 and 2.17 which show the contours of the energy density field around the crack tip as calculated from the direct approach as well as from a knowledge of the bulk stress on the surface.

Slight errors in the energy and displacement fields occur in the region of the loaded edge. However, the energy field around the crack is particularly accurate. The stress intensity factors, as calculated from a knowledge of the bulk stress alone, differ from those previously calculated by less than 1%. It thus appears that, as first postulated in [23, 29] a knowledge of the thermal emission profile and a number of strain gauge values may enable the individual stress components to be evaluated. This approach, when coupled with existing C-scan techniques, raises the possibility of the structural characterization of damage, and the nondestructive evaluation of critical damage sizes and loads.

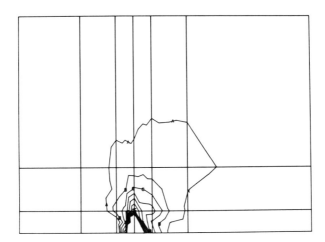

Figure 2.16. Energy density field: direct solution. A = 0.19 MPa, B = 0.28 MPa, C = 0.37 MPa, D = 0.45 MPa.

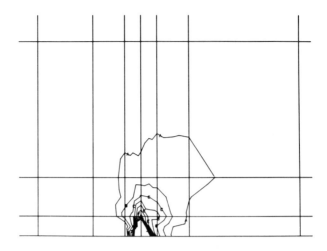

Figure 2.17. Energy density: bulk stress solution. A = 0.19 MPa, B = 0.28 MPa, C = 0.37 MPa, D = 0.45 MPa.

2.7 Conclusions

This work has shown how for a variety of practical problems, a knowledge of the energy density field can be used to assess damage severity and to develop appropriate repair stresses.

The coupling between the mechanical and thermal energies provided an assessment of energy field from the use of surface temperature measurements. However, until recently, this approach did not allow the determination of the individual stress components.

Presented is a methodology that has the potential to determine both the surface and sub-surface-stress fields. This approach therefore enables the determination of energy fields from these surface temperature measurements, thus allowing the accurate assessment of damage. A knowledge of the energy field can also be used to arrive at an optimum design.

References

[1] Sih G.C. and Chao C.K., "Failure initiation in unnotched specimens subjected to monotonic and cyclic loading", *J. Theoret. Appl. Fracture Mech.* **2** *(1)*, pp. 67–73 (1984).
[2] Sih G.C., "Mechanics and physics of energy density theory", *J. Theoret. Appl. Fracture Mech.* **4** *(3)*, pp. 157–173 (1985).
[3] Matic P. and Jolles M.I., "Fracture mechanics, strain energy density and critical issues for research and applications", Naval Research Laboratory Memorandum Report 5838, August (1986).
[4] Gillemot L.F., "Criterion of crack initiation and spreading", *Engrg. Fracture Mech. 8*, pp. 239–253 (1976).

[5] Gillemot L.F, Czoboly E., and Havas I., "Fracture mechanics applications of absorbed specific fracture energy: notch and unnotched specimens", *J. Theoret. Appl. Fracture Mech. 4*, pp. 39–45 (1985).

[6] Gao Q., Sun X.F., and Gue X.M., "Describing and predicting the high temperature LCF life with the local equivalent strain energy approach", *Proc. Internat. Conf. on Fracture and Fracture Mechanics, Shanghai, China*, pp. 381–390 (1987).

[7] Jones R., Wong A.K., Heller M., and Williams J.F., "The role of energy related parameters in life enhancement procedures", *Proc. 5th Internat. Conf. on Finite Element Methods, Melbourne*, July (1987).

[8] Molent L. and Jones R., "Stress analysis of a boron/epoxy reinforcement for the F-111C wing pivot fitting", ARL Structures Report Number 426, ARL Melbourne, Australia (1987).

[9] Badaliance R. and Dill H.D., "Compression fatigue life prediction methodology for composite structures", Vol. 1: Summary and Methodology Development', NADC-83060-60 (1982).

[10] Belgen M.H., "Infrared radiometric stress instrumentation application range study", NASA CR-1067, 1968.

[11] Cummings B., "Thermoelastic stress analysis", *Journal of Engineering, Technical File* No. 132, pp. 1–4, 1985.

[12] Stanely P. and Chan W.K., "Quantitative stress analysis means of the thermoelastic effect", *Journal of Strain Analysis*, 20 (3), pp. 129–137, 1985.

[13] Machin A.S. , Sparrow J.G.,and Stimson M.G., "Mean stress dependence of thermoelastic constant", *Journal of Strain*, **23**, pp. 27–30, 1987.

[14] Brust F.W., Nishioka T., Atluri S.N., and Nakagaki M., "Further studies on elastic plastic stable fracture utilizing the *T** integral", *Engrg. Fracture Mech*, **22**, pp. 1079–1103 (1985).

[15] Paul J., 'Evaluation of a damaged F/A-18 horizontal stabilator', Aircraft Structures Technical Memorandum 590, Aeronautical Research Laboratory, Melbourne, Australia, February (1989).

[16] Paul J. and Jones R., "Analysis and repair of impact damaged composites", Aircraft Structures Report 435, Aeronautical Research Laboratory, Melbourne, Australia, December (1988).

[17] Thomson W., (Lord Kelvin), "On the dynamical theory of heat", *Transactions Royal Society Edinburgh*, **20**, pp. 261–283, 1853.

[18] Wong A.K., Jones R., and Sparrow J.G., "Thermoelastic constant or thermoelastic parameter", *Journal of Physics and Chemistry of Solids*, **48**, pp. 749–753, 1988.

[19] Wong A.K., Dunn S.A., and Sparrow J.G. "Residual stress measurement by means of the thermoelastic effect", *Nature*, **332**, pp. 613–615, 1988.

[20] Jones R., Tay T.E., and Williams J.F., "Thermomechanical behaviour of composites", *Proceedings U.S. Army Workshop on Composite Material Response, Glasgow, Scotland 1987*, (Editors Sih, G.C., *et al*), Elsevier Publishers, London, 1988.

[21] Potter R.T. and Greaves L.J., "Application of thermoelastic stress analysis techniques to fibre composites", Optomechanical Systems Engineering, (Editor Vucobratovich, D.), *Procedings SPIE 817*, pp. 134–136, 1987.

[22] Dunn S.A., Lombardo D., and Sparrow, J.G., "The mean stress effect in metallic alloys and composites", *Proceedings at Conference on Stress and Vibration: Recent Developments in Industrial Measurement and Analysis*, London, March 1989.

[23] Heller M., Williams J.F., Dunn S. and Jones R,, "Thermomechanical analysis of composite specimens", *Journal of Composite Structures*, 11, pp. 309–324, 1989.

[24] Heller M.,"Fatigue life enhancement and damage assessment of cracked aircraft structures", PhD Thesis, University of Melbourne, 1989.

[25] Stanely P., and Chan W.K. "SPATE stress studies of plates and rings under in plane loading", *Experimental Mechanics*, pp. 360–370, December, 1986.

[26] Anderson I., "Strain survey tests of an F-111C wing pivot fitting replica specimen with a boron/epoxy doubler", Aero. Res. Lab., Struct-Tech-Memo-459, Melbourne Australia, April 1987.

[27] Grundy I., "SMOOFF – A Fortran smoothing program", Aero. Res. Lab., Struct-Tech--Memo-490, Melbourne, Australia, July 1988.

[28] Paul J., "Evaluation of a damaged F/A-18 Horizontal Stabilator", Aero. Res. Lab., Struct-Tech-Memo-503, Melbourne Australia, February 1989.

[29] Ryall T.G. and Wong A.K., "Determining stress components from thermoelastic data – A theoretical study". *Mechanics of Materials,* 7, 205–214, (1988).

[30] Waldman W., Tay T.E., and Jones R., "A penalty element formulation for calculating bulk stress", Aero Res Lab., Structures Report, Melbourne Australia, February 1989.

Effects of fillers on fracture performance of thermoplastics: strain energy density criterion

3.1 Introduction

Fillers have been widely used in commercial production of polymers to improve physical properties for specific applications. The main effects of fillers on polymers are the increase of modulus and hardness, and reduction of creep and distortion at elevated temperature [1, 2]. According to the type of reinforcement provided, fillers are often divided into three classes: one-dimensional (fibers), two-dimensional (flakes or platelets), or three-dimensional (spheres or beads).

Among the three classes of fillers, fibers and platelets are often considered to be more attractive than spheres since they result in greater stiffness and strength [3, 4]. Although fibers have advantages over the platelets in terms of efficiency of stress transfer, the latter are often chosen for reinforcement of planar structures because they can reinforce in two dimensions.

The mechanism of the reinforcement of plastics by fillers remains unclear. Many factors affect the strength of composite materials. The addition of a filler can increase or decrease the mechanical resistance of a plastic depending on the nature of the polymer and the filler. What is often assumed is that if polymers become more rigid through the addition of fillers, they also become more brittle.

Much work has been carried out to study the factors that affect the toughness of a fiber reinforced composite. Several different interactions have been identified as contributors to fracture [5–11] such as: ductilities of the fiber and the matrix, friction between fibers and matrix after the debonding process has occurred, interfacial fracture energy, fiber retraction into the matrix after brittle fracture at the crack plane, etc. When two or three-dimensional fillers are incorporated, the fracture properties of composites vary quite unpredictably. Particulate fillers can reduce [12] or enhance [13] the toughness of

G. C. Sih and E. E. Gdoutos (eds): Mechanics and Physics of Energy Density, 59–73.
© 1992 *Kluwer Academic Publishers. Printed in the Netherlands.*

plastics. Due to a concentration of stress at the matrix/filler interface, interfacial adhesion is an important factor which strongly affects the fracture toughness [14, 15].

Friedrich and Karsch [12] suggested that when a thermoplastic polymer orientates or crystallizes with strain, it exhibits poor bonding with fillers. Consequently, it is able to separate and deform away from the filler surface (to stretch and to work harden). The matrix stretching generally requires a large amount of energy in a fracture process, whereas the energy required for the formation of voids and cracking of the interface between the polymer and the filler is considerably smaller. Consequently, the addition of fillers usually results in a more or less continuous and drastic reduction of the fracture properties. However, it has been shown [16, 17] that good adhesion does not necessarily lead to an increase in crack growth resistance but the interfacial strength can affect the mechanism of plastic deformation of the matrix. Consequently, the nature of the polymer itself also plays an important role in determining the effect of a given filler on mechanical properties.

In addition to the complexity of the fracture process in multiphase materials, there is much controversy in the characterization and prediction of fracture. This paper presents the experimental results on the fracture characterization of filled thermoplastics. The emphasis is put on verification of the validity of different fracture criteria. The effects of two- and three-dimensional fillers on the fracture performance of a representative thermoplastic polypropylene, are discussed in terms of the strain energy density criterion.

3.2   Experimental consideration

An injection grade of polypropylene, Profax 6301, from Hercules was chosen for this study. The glass flakes were a commercial grade (Hammermilled Flakeglass 1/64"), supplied by Owens Corning Fiberglass. The average flake diameter, determined by sieving, was about 120 μm. Two commercial grades of phlogopite mica, supplied by Marietta Resources International, were selected; an untreated (200 HK) and a silane treated (200 NP) grade. A description of these grades is given in [16, 18]. The glass beads were obtained from Potters Industries, Laprairie, Quebec, Grade No. 1922. The diameter was between 150 and 200 μm.

Polypropylene powder containing different mica and glass-flake concentrations were dry blended in a Welex high intensity mixer, Model ESK-8C. Sample plates of 30 mm by 90 mm and of thicknesses of 3 mm and 6 mm were prepared by injection molding. Double-edge-notch and three-point-bend samples were cut out from these plates. The initial crack in the samples was created in two steps: a prenotch was first made by a saw cut and the exact final crack length was created by forcing a razor blade into the specimen using a special jig on a laboratory press.

3.3 Fracture analysis

The effect of filler on fracture behavior of polypropylene is schematically represented in Figure 3.1. In the presence of a crack, virgin polypropylene exhibits a purely brittle behavior. In three-point-bending, fracture occurs at a constant deflection with the aid of the energy provided by the elastic strain energy stored in the sample. With the addition of mica, the fracture behavior of mica-reinforced polypropylene becomes more and more ductile. Up to 5 percent by weight of mica, the unstable fracture (at constant displacement) still occurs, but the amounts of plastic deformation and stable crack growth are increased with mica content. The plastic deformation accompanied by stable crack propagation is represented by the non-linear region of the load versus displacement curve. Since the area under these curves increases, the energy absorbed by the specimen to fracture increases with mica content, all the specimens tested having the same crack length and dimensions. Above 10 percent, no unstable fracture was observed at the cross-head speed of 10 mm/min. The crack grows with continuous supply of energy from the external load (with continuous increase in displacement until complete fracture). The amount of energy absorbed by the specimen to break, which is represented by the area under the load deflection diagram, continues to augment up to 20

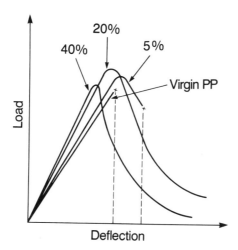

Figure 3.1. Schematic drawing showing load-deflection curves of the three-point bending fracture tests at different mica concentrations. X represents the point at unstable, brittle fracture.

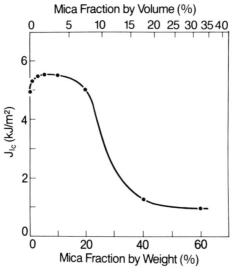

Figure 3.2. *Fracture energy at crack initiation (J_{1c}) as a function of mica concentration.*

percent mica. Above 20 percent, the stable fracture still occurs, but the area under the load deflection curve decreases as the mica content is increased.

Crack specimen. Standard methods of fracture test recommend the use of crack initiation for the measurement of the material's toughness. Figure 3.2 shows the variation of such a toughness parameter with mica concentration in polypropylene. The reported parameter is the critical value of the contour (J) integral, determined from three-point-bend specimens according to the ASTM recommended standard procedure [19]. The result confirms the previously reported statement that the addition of fillers leads to a more or less drastic reduction in the fracture toughness of polypropylene [12]. However, from the above observation, it is clear that the measured J_{1c} values do not reflect the fracture behavior of the composite. Depending on mica concentration, stable or unstable fracture can occur at identical mechanical conditions. This behavior is of prime importance in designing with plastics because one of the major concerns with plastic parts is the shattering problem which can cause injury to people. Analysis of the shattering phenomenon in practice suggests that the velocities of the projected debris are much higher than the loading rate. This could only be possible if the kinetic energy of the medium before separation into pieces is very high. In order to acquire this kinetic energy, the crack propagation velocity must be large. The crack growth has therefore to be maintained with the aid of the elastic energy stored in the structure instead of the external forces. Instability under a displacement control condition (instead of load control condition) must then be used to predict shattering. In an unstable crack propagation at constant deflection in the three-point-bending tests, the two halves of the sample fly away after fracture. This obviously corresponds to the shattering condition of plastic parts. A valid fracture criterion should therefore be able to predict such an instability.

At present, unstable crack propagation is explained from a global analysis of the structure. Instability is assumed to occur when the strain energy release rate exceeds the fracture energy of the material. In order to verify the validity of this criterion, the locus of fracture points on the load-deflection diagram of the three-point-bend can be used. The expression for the stress intensity factor for three-point-bend samples is given by [20]:

$$K_{1c} = \frac{3}{2} \frac{P_c L}{B D^{3/2}} \left(\frac{a}{D}\right)^{1/2} Y \tag{3.1}$$

in which L is the span, B is the thickness, D the specimen's width, a is the crack length, P_c is the load at fracture and Y is a calibration factor defined in [20].

From the definitions of the critical strain energy release rate:

$$G_c = -\frac{1}{2} P_c^2 \frac{dC}{dA} \tag{3.2}$$

and also

$$G_c = \frac{(1-v^2)K_c^2}{E}, \tag{3.3}$$

in which E is the Young modulus, A the crack area and v the Poisson ratio. The loading path on the load-deflection diagram for a given crack length can be defined by the specimen's compliance:

$$C = C_0 + \frac{9}{2} \frac{1-v^2}{EB} \left(\frac{L}{D}\right)^2 \int_0^{a/D} \left(\frac{a}{D}\right) Y^2 d\left(\frac{a}{D}\right), \tag{3.4}$$

where C_0 is the specimen's compliance for $a = 0$. From the above expressions, at a constant value of G_c, the corresponding value of P_c can be determined.

Fracture data. Figure 3.3 shows the locus of fracture points for the case $G_c = 1$ kJ/m², $E = 2000$ MPa, $B = 6$ mm, $D = 13$ mm, and $L/D = 4$. The left-hand side of this locus corresponds to the non-fracture region and the right-hand side corresponds to the fracture region. According to this curve, under a displacement control condition, fracture can take place by two different modes, depending on the crack length in the sample. If $a/D < 0.35$, when the loading path meets the locus of fracture points, crack initiation occurs. The crack growth results in a load drop and the loading path moves into the

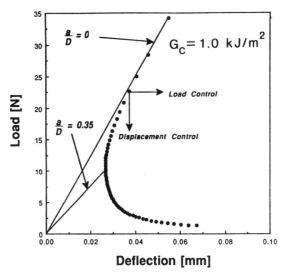

Figure 3.3. Locus of fracture points for the three-point-bend samples used with a fracture energy, $G_c = 1$ kJ/m².

fracture region at the same displacement. In this case, fracture should be unstable and eventually followed by a crack arrest when the loading path meets the lower portion of the locus of fracture points. In contrast to this, when $a/D \geqslant 0.35$, any crack propagation moves the loading path into the non-fracture region at a constant displacement. Fracture can only occur if the displacement is increased. The fracture is thus stable according to the constant G_c criterion. The experimental results actually revealed that fracture does not take place as predicted by this scheme. As discussed above, at a given crack length, stable or unstable fracture can take place depending on the filler concentration. For instance, pure polypropylene always breaks in an unstable manner even if $a/D > 0.35$. Stable or unstable crack propagations have been found to only depend on the material behavior. Between 0 and 10% mica concentration by weight, stable fracture takes place before the occurrence of instability. The amount of stable crack growth also increases with increasing filler content.

3.4 Strain energy density criterion

The fracture behavior can be explained in terms of the strain energy density criterion [21–27]. The theory postulates that for any nontrivial stress state, there exists at least one maximum and one minimum of the strain energy density function, dW/dV, referred to a coordinate system at each point in a medium. Among many of the minima of dW/dV, the maximum of $(dW/dV)_{min}$ is assumed to coincide with macrocrack growth and the largest of $(dW/dV)_{max}$

with the direction of maximum permanent deformation. The former can be associated with excessive dilatation while the latter with excessive distortion. Fracture is assumed to occur when the maximum of $(dW/dV)_{min}$ reaches a critical value at a certain distance ahead of the crack tip. This distance determines the magnitude of the incremental crack growth and fracture becomes unstable when the crack growth increment reaches a critical value such that:

$$\left(\frac{dW}{dV}\right)_{min} = \frac{S_1}{r_1} = \frac{S_2}{r_2} = \cdots = \frac{S_i}{r_i} = \cdots \frac{S_c}{r_c} \rightarrow \left(\frac{dW}{dV}\right)_c \qquad (3.5)$$

in which S_i is the strain energy density factor and r_i is the incremental crack growth. Instability of fracture can therefore be predicted by any two of the three material parameters $(dW/dV)_c$, S_c and r_c.

Fracture surface. The examination of the fracture surfaces of various thermoplastics revealed that in the case of unstable fracture, there is always a region of formation and coalescence of voids in front of the crack tip and unstable fracture is initiated from this region. Figure 3.4 shows the fracture surface of pure polypropylene. It can be clearly seen that the region of void coalescence in front of the initial crack constitutes the initiation of fracture. The initiation point in unstable fracture is also known as the fracture nucleus. This observation suggests that fracture instability is governed by a local material parameter. The location of fracture initiation found at a certain

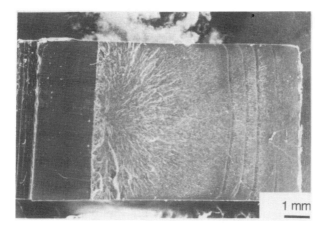

Figure 3.4. Fracture surface of pure polypropylene showing the region of fracture initiation in front of the crack tip.

distance ahead of the crack tip is in agreement with the assumption in the strain energy density theory. As discussed above, the theory postulates that crack propagation is incremental and instability will occur if the increment of crack growth reaches a critical value. In a purely brittle, unstable fracture mode such as in the case of pure polypropylene, it could be postulated that the first crack-growth increment immediately reaches the critical value, r_c, of the material. On the other hand, a completely stable fracture, such as that which occurs in polypropylene containing a high filler concentration, could be explained by the fact that the last increment of crack growth (at the complete separation of the sample) has not reached a critical value. However, if the size of the sample is large enough (which is the case for a real structure) to allow the crack growth increment to attain the critical value, unstable fracture can occur. According to the strain energy density theory, r_c is assumed to be a material characteristic. In order to verify the existence of this parameter, three-point-bending tests were carried out on samples containing the same notch length but different notch tip radii. By changing the notch tip radius, the profile of the strain energy density in front of the notch would be changed as shown schematically in Figure 3.5. As the notch radius increases, the strain energy density will reach the same value at a larger distance ahead of the notch tip. The increment of crack growth will therefore augment. If this distance attains the critical value, r_c, and if this value exists, unstable fracture should occur. The experimental results show that when the notch tip radius is large enough, fracture becomes unstable, as shown schematically in Figure 3.6. The result confirms the existence of the critical value, r_c, characterizing the instability of fracture in the material. Figure 3.7 shows the variation of the critical notch tip radius at which fracture becomes unstable, as a function of mica concentration. When mica content increases the critical notch tip radius becomes larger, suggesting that mica addition results in an increase in the

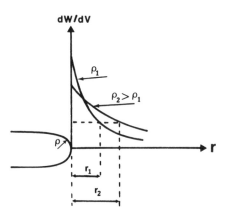

Figure 3.5.   Effect of notch tip radius on the increment of crack growth.

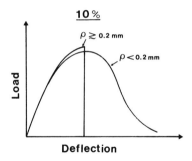

Figure 3.6. Effect of notch-tip radius on fracture behavior of glass-flake reinforced polypropylene (10 wt% glass flakes).

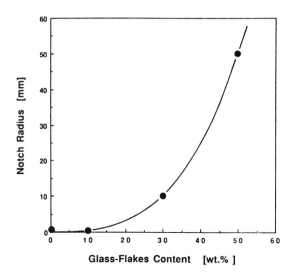

Figure 3.7. Critical notch-tip radius at unstable fracture as a function of glass flake content.

critical increment of crack growth at instability. On the macroscopic scale, the tensile test shows that filler addition results in a reduction in the critical strain energy density, $(dW/dV)_c$. Figures 3.8 and 3.9 show the variation of $(dW/dV)_c$ as a function of filler concentration for the cases of mica and glass flakes respectively. Globally the composites become more and more brittle with filler

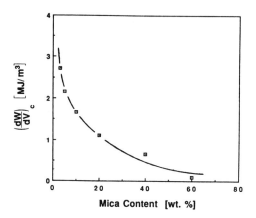

Figure 3.8.   Effect of mica content on the critical strain energy density of mica/polypropylene composite.

addition. However, as discussed above, on the microscopic level, fillers augment the critical increment of crack growth at instability. The composites would therefore be less prone to shatter in the presence of a crack. Since it has long been recognized that defects exist in most structures, filled thermoplastics would appear less brittle in service.

It is worth mentioning that with the sizes of the specimen used, the complete fracture process could be observed at low filler contents. The crack grows in a stable manner over a certain distance and suddenly propagates unstably.

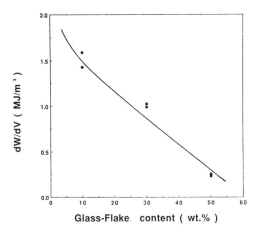

Figure 3.9.   Effect of glass-flake content on the critical strain energy density of glass flakes/ polypropylene composite.

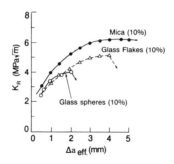

Figure 3.10. Critical stress intensity factor, K_R, as a function of crack extension for different types of filler.

The period of stable crack growth also increases with filler concentration. The variation of the strain energy density factor during the stable crack propagation period can be determined by measuring the variation of the critical stress intensity factor, K_R, since these parameters are directly related [22]. Figure 3.10 shows the variation of K_R as a function of crack extension for different types of filler. The experimental results also suggest that the maximum values of the critical stress intensity factor at instability, K_{Rmax}, remain relatively constant regardless of the initial crack length. Since K_{Rmax} is related to S_c, the results confirm the assumption in equation (3.5) on the material parameters characterizing unstable fracture.

Interfacial strength. Another major factor which controls the toughness of composites is the interfacial strength. At present, it is generally assumed that a better adhesion between the two phases would result in a tougher composite since any increase in the adhesion energy will add to the fracture energy of the material. In fact, this assumption is confirmed by the measurements of the classical single fracture parameter. For instance in the case of mica/polypropylene composite, the use of a coupling agent such as silane results in a higher J_{1c} value as shown in Figure 3.11. Nevertheless, in practice, it has often been reported that a better matrix/filler interface results in materials with lower impact resistance. As demonstrated above, fracture cannot be properly characterized by a single parameter in classical concepts. In the presence of a defect such as a crack, the critical increment of crack growth prior to instability should also be taken into account in order to completely describe the fracture process. The effect of filler treatment on the r_c value can be demonstrated by determining the critical notch tip radius leading to unstable fracture. Table 3.1 shows the limit notch-tip-radii, ρ_c, for the untreated and silane-treated mica/polypropylene composites. With a better interface, the ρ_c value becomes smaller, suggesting a lower value of r_c in the composite. Consequently, fracture would be less stable and the composite could exhibit

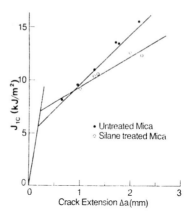

Figure 3.11. Effect of coupling agent on *J*-integral measurement (20 wt% mica).

Table 3.1. Effect of silane treatment on critical strain energy density and critical notch-tip radius at unstable fracture.

	PP / 30% GLASS FLAKES	
	$\left(\dfrac{\mathrm{d}W}{\mathrm{d}V}\right)_c$ (MJ/m³)	Critical Notch Radius (mm)
Untreated	1	10
Silane treated	0.8	3

a poorer impact resistance when a coupling agent is used. The physical significance of this effect can also be interpreted in terms of damage development on the microscopic level. It has been suggested [27] that the increment of crack growth is controlled by the relative critical strain energy density of the damaged material, $(\mathrm{d}W/\mathrm{d}V)_c^*$, which takes into account the influence of smaller scale damage on the local material toughness. According to the proposed scheme, the effect of damage on the crack growth increment is presented in Figure 3.12. The amount of crack propagation is assumed to be determined by the intersection between $(\mathrm{d}W/\mathrm{d}V)_c^*$ and $\mathrm{d}W/\mathrm{d}V$. When the interfacial adhesion is improved, damage is reduced and the relative critical strain energy would be higher. The intersection with the singular strain energy field would be found at a shorter distance ahead of the crack tip, resulting in a smaller increment of crack growth. Examination of the fracture surfaces

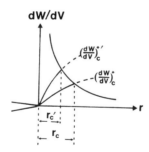

Figure 3.12. Effect of microscopic damage on increment of crack growth.

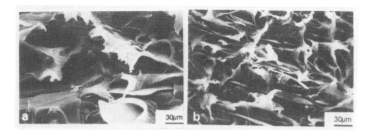

Figure 3.13. Micrograph of fracture surface of polypropylene reinforced by 20 percent of: (a) untreated mica and (b) silane-treated mica.

(Figure 3.13) also shows that in the case of treated mica, the matrix stretching is less significant than that in the case of untreated mica. The local material toughness parameters are therefore in agreement with previous observations on the effect of coupling agents on the impact resistance of filled thermoplastics.

3.5 Conclusion

It has been shown in this work that classical fracture mechanics concepts do not properly describe the fracture behavior of filled thermoplastics. The weakness of these concepts remains in the inability to predict stable crack growth followed by fracture instability. Spurious conclusions on the material toughness could be made if the complete fracture process is not considered, especially when these parameters are used to evaluate the material resistance to shattering. The strain energy density theory has been found to provide a more appropriate criterion for fracture. On the continuum scale, the addition of fillers leads to a reduction of the critical strain energy density of most thermoplastics. However, on the microscopic level fillers strongly increase the amount of crack growth increment. Fracture becomes more and more stable with increasing filler concentration. The critical increment of crack growth prior to fracture instability increases with filler content and depends on the smaller scale damage, induced by the debonding at the matrix/filler interface. Microscopic observation of the fracture surfaces suggests that unstable fracture is initiated at a certain distance ahead of the main crack, which is in agreement with the strain energy density theory. Measurements made on materials exhibiting purely stable fracture, containing different notch tip radii, also confirm the existence of the critical increment of crack growth as a material characteristic. The smaller scale damage, induced by the debonding between the matrix and fillers interface, is responsible for the augmentation of the crack growth increment and consequently for a more stable fracture.

Improvement of the filler/matrix adhesion with coupling agents reduces the microscopic damage and the critical incremental crack growth prior to instability. The results confirm the role of the smaller scale damage predicted by the strain energy density concept and are in agreement with a negative effect of coupling agents on the impact resistance observed in practice.

References

[1] Richardson, M.O.W., "Polymer engineering composites", Applied Science Publishers Ltd, London, 1977.
[2] Woodhams, R.T. and Xanthos, M. "Handbook of fillers and reinforcements for plastics", H.S. Katz, J.V. Milewshi, Eds., p. 333, Van Nostrand Reinhold Co., New York, 1978.
[3] Richardson, G.C. and Sauer, J.A., "Effect of reinforcement type on the mechanical properties of polypropylene composites", *Polym. Eng. Sci.,* 16, p. 252, 1976.
[4] Busigin, C., Martinez, G.M., Woodhams, R.T., and Lahtinen, R., "Factors affecting the mechanical properties of mica-filled polypropylenes", *Polym. Eng. Sci.,* 23, p. 766, 1983.
[5] Helfet, J.L. and Harris, B., "Fracture toughness of composites reinforced with discontinuous fibres", *J. Mater. Sci.,* 7, p. 494, 1972.
[6] Cooper, G.A., "The fracture toughness of composites reinforced with weakened fibres", *J. Mater. Sci.,* 5, p. 645, 1970.
[7] Beaumont, P.W. and Harris, B., "The energy of crack propagation in carbon fiberrein-forced resin systems", *J. Mater. Sci.,* 7, p. 1265, 1972.
[8] Pigott, M.R., "The effect of aspect ratio on toughness in composites, *J. Mater. Sci.,* 9, p. 494, 1974.
[9] Marston, T.U., Atkins, A.G., and Felbeck, D.K., "Interfacial fracture energy and the toughness of composites", *J. Mater. Sci.,* 9, p. 447, 1974.
[10] Barris, B., Morley, J., and Philips, D.C., "Fracture mechanisms in glass-reinforced plastics", *J. Mater. Sci.,* 10, p. 2050, 1975.
[11] Miwa, M., Ohsawa, T., and Tsuji, N., "Temperature dependence of impact fracture energies of short glass fiber – epoxy and unsaturated polyester composites", *J. Appl. Polym. Sci.,* 12, p. 1679, 1979.
[12] Friedrich, K.J. and Karsch, U.A., "Failure processes in particulate filled polypropylene", *J. Mater. Sci.,* 16, p. 2167, 1981.

[13] Bramuzzo, M., Savadori, A., and Bacci, D., "Polypropylene analysis of impact strength", *Polym. Compos.*, **6**, p. 1, 1985. *Polypropylene composites:fracture mechanics analysis of impact strenqth.*

[14] Leidner, J. and Woodhams, R.T., "The strength of polymeric composites containing spherical fillers", *J. Appl. Polym. Sci.*, **18**, p. 1639, 1974.

[15] Radosta, J.A., "Impact and flexural modulus behavior of calcium, carbonate and talc filled polyolefins", SPE Tech. Papers, **22**, p. 465, 1976.

[16] Vu-Khanh, T. Fisa, B., "Fracture behavior of mica-reinforced polypropylene: effects of coupling agent, flake orientation, and degradation", *Polym. Compos.*, **7**, p. 220, 1986.

[17] Abate, G.F. and Heikens, D., "Polymer – filler interaction in polystyrene filled with coated glass spheres: 2. Crazing modes at different adhesive strength in impact experiments", *Polym. Commun.*, **24**, p. 342, 1983.

[18] Vu-Khanh, T., Sanschagrin, B., and Fisa, B., "Fracture of mica-reinforced polypropylene: mica concentration effect", *Polym. Compos.*, **6**, p. 249, 1985.

[19] E 813 Standard test method for J_{Ic}, a measure of fracture toughness, *Annual Book of ASTM Standards, 1983.*

[20] E 399 Standard test method for plane strain fracture toughness of metallic materials, *Annual Book of ASTM Standards, 1983.*

[21] Sih, G.C., "Some basic problems in fracture mechanics and new concepts", *Eng. Fract. Mech.*, **5**, p. 365, 1983.

[22] Sih, G.C. and Macdonald, B., "Fracture mechanics applied to engineering problems – strain energy density fracture criterion", *Eng. Fract. Mech.*, **6**, p. 361, 1974.

[23] Sih, G.C. and Madenci, E., "Crack growth resistance characterized by the strain energy density function", *Eng. Fract. Mech.*, **18**, p. 1159, 1983.

[24] Gdoutos, E.E. and Sih, G.C., "Crack growth characteristics influenced by load time record", *J. Theoretical and Appl. Fract. Mech.*, **3**, p. 95, 1984.

[25] Sih, G.C. and Chen, C., "Non-self-similar crack growth in elastic-plastic finite thickness plate", *J. Theoretical and Appl. Fract. Mech.*, **3**, p. 125, 1985.

[26] Sih, G.C., "Mechanics and physics of energy density theory", *J. Theoretical and Appl. Fract. Mech.*, **4**, p. 157, 1985.

[27] Matic, P. and Sih, G.C., "A pseudo-elastic material damage model, Part I: Strain energy density criterion", *Theoretical and Appl. Fract. Mech.*, **3**, p. 193, 1985.

Strain energy density criterion applied to characterize damage in metal alloys

4.1 Introduction

Damage accumulation in metal alloys is a time-dependent process that is highly nonlinear and deviates substantially from thermodynamic equilibrium. Phase transition of metals would involve change in the material characteristics at different scale levels. A synergetic consideration of different properties would be required. It is the objective of this work to show that the strain energy density criterion of Sih can be applied to evaluate the damage thresholds in the hierarchy of phase changes in metals. The three phases to be considered are the initial crystalline state; the quasi-amorphous state; and the damage state involving micropores and microcracks in the bulk. Distinct character of the latter two states can be identified with their critical strain energy densities. The transition from microdamage to macrodamage prior to complete material separation can thus be assessed. Applied is the concept of fractals for defining the ratio between the size of the self-similar zone before failure and critical size of the macrocracks along the path of the prospective crack growth. This is accomplished by application of the Sih theory which accounts for the transition from subcritical crack growth to rapid fracture. Characterized are the resistance to plastic strain governed by the material yield strength in the initial state; resistance to crack formation controlled by the critical strain energy density; and resistance to crack propagation associated with the driving energy of the crack.

4.2 Strain energy density criterion

The energy density concept advanced by Sih [1, 2] is most informative because it simultaneously addresses crack initiation, stable growth and onset of rapid fracture. Time dependent changes in the strain energy density function dW/dV can be obtained as the area under the true stress and true strain curve given by [3]:

G. C. Sih and E. E. Gdoutos (eds): Mechanics and Physics of Energy Density, 75–85.
© 1992 *Kluwer Academic Publishers. Printed in the Netherlands.*

$$\frac{dW}{dV} = \int_0^{\varepsilon_{ij}} \sigma_{ij} \, d\varepsilon_{ij} + F(\Delta T, \Delta C). \tag{4.1}$$

In equation (4.1), σ_{ij} and ε_{ij} are, respectively, the stress and strain components while ΔT and ΔC are changes in the temperature and moisture concentration in the medium. Since dW/dV is time dependent and is different for each material element, the evolution of damage of the system can be completely described in a unique fashion without ambiguity [1]. Refering to the true stress and true strain curve in Figure 4.1, the region $oypq$ corresponds to the energy density $(dW/dV)_p$ dissipated at the instance p. The region $pp'q$ is

$$\left(\frac{dW}{dV}\right)^* = \frac{dW}{dV} - \left(\frac{dW}{dV}\right)_p \tag{4.2}$$

represents the available energy density to do further damage in the material. As the stress and strain change with time, dW/dV in each local element alters

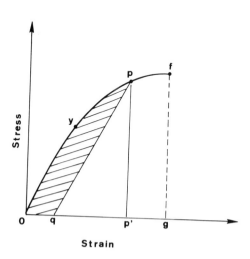

Figure 4.1. Schematic of time stress versus true strain.

accordingly. The critical value of dW/dV or $(dW/dV)_c$, however, is characteristic of the material [1, 3]; it corresponds to the total region of g in Figure 4.1. In what follows, the symbol $\mathscr{W}$ will be used to denote dW/dV. As Sih [1] has pointed out, test specimens under ordinary laboratory conditions should be regarded as an open system where energy exchange prevails between the system and its surroundings.

4.3 Thermal/mechanical interaction in solids

When a solid is stressed mechanically up to its ultimate plastic strain, the local energy state is similar to that of the crystalline phase at the melting point [4, 5]. Use can thus be made of the specific thermodynamic potential of the system at the melting point which corresponds to the terminal values of the Gibbs function for the crystalline and liquid states. These properties apply to open systems [6].

Dilatation and distortion. According to Sih [1], the elements in a solid, when strained, experiences both dilatation and distortion. The total strain energy density $\mathscr{W}$ can thus be divided into two components: one related to dilatation, $\mathscr{W}^v$, and one to distortion, say $\mathscr{W}^d$. Their corresponding critical values shall be denoted by $\mathscr{W}^v_c$ and $\mathscr{W}^d_c$. It shall be further assumed in this work that both $\mathscr{W}^v_c$, $\mathscr{W}^d_c$ are material constants independent of the ways with which energy is applied to the system. Moreover, they can be evaluated from the energy state corresponding to melting [4, 7]. Refer to the thermodynamic strength criteria explained in [8, 9].

The internal energy density arising from heating takes the form

$$H_{TS} = \int_{T_c}^{T_s} C_p(T)\,dT \tag{4.3}$$

with C_p being the heat capacity at constant pressure. For temperature T below T_c, heating due to thermal oscillation can be neglected in the internal energy density expression. Energy density associated with shape change is the same as the latent heat of melting L_m. Division of $\mathscr{W}$ into $\mathscr{W}^d$ and $\mathscr{W}^v$ is equivalent to the decomposition of the stress tensor into its deviatoric and hydrostatic components. A description of the state of affairs ahead of a macrocrack is shown in Figure 4.2 where elements are damaged according to dilatation and distortion at the microscopic and macroscopic scale level.

Critical stresses. If σ_c and τ_c stand for the critical normal and shear stress, respectively, then $\mathscr{W}^v_c$ and $\mathscr{W}^d_c$ can be written as

$$\mathscr{W}^v_c = \frac{1}{2E}\,\sigma_c^2, \qquad \mathscr{W}^d_c = \frac{1}{2G}\,\tau_c^2, \tag{4.4}$$

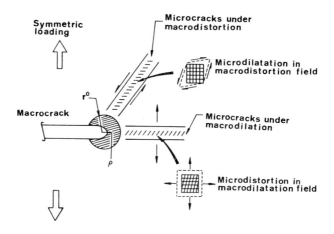

Figure 4.2. Dilatation and distortion of elements near macrocrack tip [1].

where E and G are the Young's and shear modulus of elasticity. For a three-dimensional stress state with $\sigma_j (j = 1, 2, 3)$ being the principal stresses, σ_c in equation (4.4) takes the form of the first stress invariant:

$$\sigma_c = \frac{1}{3}\,(\sigma_1 + \sigma_2 + \sigma_3). \tag{4.5}$$

Similarly, τ_c can be the octahedral shear stress which is related directly to the distortion energy density component. In view of equations (4.4), the following ratio is obtained,

$$\frac{\tau_c}{\sigma_c} = \left[\frac{G}{E}\cdot\frac{\mathcal{W}_c^d}{\mathcal{W}_c^v}\right]^{1/2}. \tag{4.6}$$

Phase change. Illustrated in Figure 4.3 is the transition of stage I to II then to III which correspond, respectively, to the crystalline phase, no-shear resistance phase and quasi-amorphous phase. At stage II, $\mathcal{W}_c^v = H_{TS}$ while $\mathcal{W}_c^d = L_m$ at stage III. Under these considerations, equation (4.6) becomes

$$\frac{\tau_c}{\sigma_c} = \sqrt{\bar{\Delta}} \tag{4.7}$$

in which Δ stands for

$$\Delta = \frac{G}{E}\cdot\frac{L_m}{H_{TS}} \tag{4.8}$$

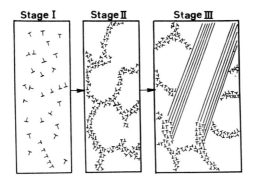

Figure 4.3. Phase transitions; Stage I – Crystalline state; Stage II – no-shear resistant state; and Stage III – Quasi-amorphous phase.

Equation (4.8) has been used to calculate Δ at T_c equal to the room temperature and found to be close to those for the alloys in [7, 10, 11]. Alloying tends to increase T_c above room temperature in situations where a noticeable increase in dilatation is observed. Alternatively, the ratio $\mathcal{W}_c^d / \mathcal{W}_c^v$ can be expressed as

$$\frac{\mathcal{W}_c^d}{\mathcal{W}_c^v} = 2(1 + v)\,\Delta. \tag{4.9}$$

Damage at the microscopic scale level can thus be determined from the ratio of the critical distortional energy density component to the critical dilatational energy density component. In order to associate the damage at the microscopic level to that at the macroscopic level, Sih's concept on the strain energy density function will be employed where the critical value $\mathcal{W}_c$ can be used to denote the onset of macrodamage.

4.4 Damage characterization

Since the strain energy density theory can be applied at any scale level providing that the quantities $\mathcal{W}$, $\mathcal{W}^d$ and $\mathcal{W}^v$ are evaluated accordingly, it will be used to establish the transition of subcritical microcrack growth to the onset macrocrack instability. The microcracks are assumed to grow intermittently such that their size are proportional to the strain energy density factor S. That is, the ratio S/r_j^0 remains constant along every new crack front satisfying the condition $\mathcal{W}_c$ constant. The subcritical growth condition can be stated as [2]:

$$\mathcal{W}_c = \frac{S_1}{r_1^0} = \frac{S_2}{r_2^0} = \cdots = \frac{S_j}{r_j^0} = \cdots = \frac{S_c}{r_c^0} = \text{const.} \tag{4.10}$$

The incremental advancement of the microcrack is r^0_j where $j = 1, 2, ..., n$. At instability, where the microcracks have coalesced, r^0_c represents a macroscopic fracture ligament. The corresponding strain energy density factor is S_c and can be related to the conventional fracture toughness value K_{1c} [2]:

$$S_c = \frac{(1 + v)(1 - 2v) K^2_{1c}}{2\pi E}.$$
(4.11)

Scale factor. In view of equation (4.10) and (4.11), $\mathcal{W}_c$ can also be solved in terms of K_{1c}. Solving for r^0_c, it is found that

$$r^0_c = \frac{(1 + v)(1 - 2v)}{2\pi E \mathcal{W}_c} K^2_{1c}.$$
(4.12)

Let r^{max}_c be the maximum size of the zone prior to fracture as shown in Figure 4.4 for Mode I loading. It can be determined from [12]:

$$r^{max}_c = \frac{1}{2} \left[\frac{K^{max}_{IR}}{\sigma_{ys}}\right]^2,$$
(4.13)

where σ_{ys} is the yield stress and K^{max}_{IR} corresponds to the stress intensity factor as damage spread over beyond r^0_c ahead of the macrocrack in Figure 4.4. The ratio r^{max}_c / r^0_c may thus be interpreted as a scale factor, say Λ^c:

$$\Lambda_c = \frac{\lambda^* \mathcal{W}_c}{(K_{1c} \sigma_{ys})^2}$$
(4.14)

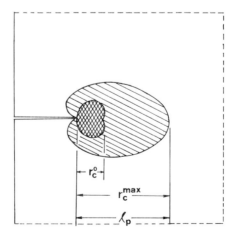

Figure 4.4. Definition of scale factor.

in which λ^* stands for

$$\lambda^* = \frac{E(K_{IR}^{max})^2}{(1 + v)(1 - 2v)}. \tag{4.15}$$

Note that the damage zone of the material is r_c^{max} while the size of the microdamage zone is r_c^0.

General characterization. In view of the foregoing treatment, the K_{1c}, W_c and σ_{ys} can in general be related to a single parameter*

$$p^* = \frac{(K_{1c}\,\sigma_{ys})^2}{W_c}. \tag{4.16}$$

The validity of Λ^c and p^* can be exhibited for different grades of steels. A plot of log Λ^c versus log p^* is shown in Figure 4.5 which reveals that all data fall on a straight line. In what follows, the concept of fractals will be applied treating the strained solid as a synergetic system [13] to establish a relation between damage at the microscopic and macroscopic scale level [14].

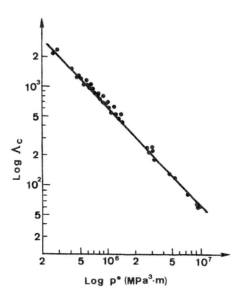

Figure 4.5. Scale factor Λ_c as a function of p^*.

---

* Alternatively, the critical energy release rate G_{1c} may be grouped with σ_{ys}^2 and W_c as $G_{1c}\,\sigma_{ys}^2/W_c$.

4.5 Transition of Micro- to Macrodamage

Relations between parameters referred to microdamage and macrodamage can be established by making use of concepts developed on fractals as self-modelling objects. By self-modelling [15], it is meant that there exists a function f that reproduces itself:

$$f^n = \frac{1}{N},\tag{4.17}$$

where η is a fractal dimension such that $0 \leqslant \eta \leqslant 3$. The parameter η can be correlated to the roughness of the failure surface [16]. Fracture surface varies in the range of $2 \leqslant \eta \leqslant 3$ while $1 \leqslant \eta \leqslant 2$ corresponds to a broken line or perforated surface. In the latter case, $\eta = \log A / \log D$ where A is the area and D the perimeter of the self-modelling object.

Unlike mathematical sets, self-modelling of natural objects is limited by the scale [17]. Suppose that the characteristic boundaries coincide with subcritical crack growth loading to fracture. Consider the zonal clusters as the self-modelling objects such that stress relaxation may occur on account of non-equilibrium phase transition, say involving the change of the crystalline phase to a quasi-amorphous one. Critical damage may thus correspond to the clusters attaining their critical density. Self-modelling growth may be regarded as an avalanche that occurs when the cluster size ahead of the crack reaches its critical r_c^0. In this case, the scale factor as referred to l_p in Figure 4.4 may be written as

$$\Lambda_c = \frac{l_p}{r_c^0} = \frac{r_c^{\max}}{r_c^0},\tag{4.18}$$

The situation when $l_p = r_c^{\max}$ is shown in Figure 4.6(a) whereas the case of $l_p = r_c^{\max}$. Δ is given in Figure 4.6(b). Since the maximum ligament of macrocrack extension is limited to the maximum size of the damage zone $r_c^{\max}$, the scale factor Λ_c may be used as N in equation (4.17) to characterize the number of self modelling objects taken as zone with size r_c^0 being controlled by the condition $\tau_c / \sigma_c = \text{const.}$, Δ may be taken as f in equation (4.17) and hence

$$\Delta^n = \frac{1}{\Lambda^c} = \frac{p^*}{\lambda^*}.\tag{4.19}$$

Experimental data have been obtained for steels with yield stresses in the range $200 \text{ MPa} \leqslant \sigma_{ys} \leqslant 1150 \text{ MPa}$ where $v = 0.28$, $E = 21 \times 10^4 \text{ MPa}$, $K_{IR}^{\max} = 40.3$ $\text{MPa}\sqrt{m}$ and $\Delta = 0.11$. Equation (4.19) were found to fit well for $2 \leqslant \eta \leqslant 3$ at $8.4 \times 10^5 \text{ MPa}^3 m \leqslant p^* \leqslant 7.6 \times 10^6 \text{ MPa}^3 m$. When $p^* \leqslant 8.4 \times 10^5 \text{ MPA}\sqrt{m}$, the test data best fit the expression

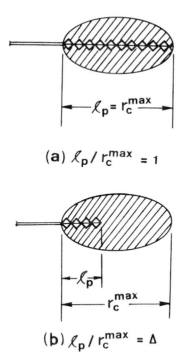

Figure 4.6. Spread of subcritical microcracks in damage zone ahead of macrocrack.

$$\Delta^{\eta + 1} = \frac{p^*}{\lambda^*},$$

(4.20)

In general, both equations (4.19) and (4.20) can be written as

$$\Delta^{\eta + m} = \frac{(1 + v)(1 - 2v)\sigma_{ys}^2}{E\mathcal{W}_c}\left[\frac{K_{1c}}{K_{1R}^{max}}\right]^2 = \frac{p^*}{\lambda^*}$$

(4.21)

such that $m = 0$ corresponds to equation (4.19) with $l_p = r_c^{max}$ and $m = 1$ to equation (4.20) with $l_p = r_c^{max} \cdot \Delta$. Plots of η versus log p^* are displayed in Figure 4.7 where curves 1 and 2 refer, respectively, to the situation in Figures 4.6(a) and 4.6(b). The range $0 \leqslant m \leqslant 1$ corresponds to failure when the region of microcrack damage is less than or at most equal to the maximum size of damage zone r_c^{max} prior to macrocrack growth. Within the range $-1 \leqslant m \leqslant 1$, microdamage is Δ times r_c^{max}. The full range of l_p corresponding to p^* is covered in Figure 4.7.

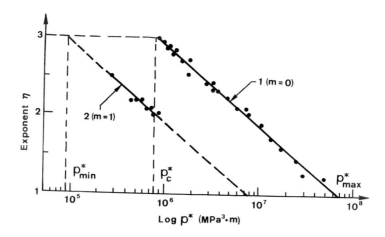

Figure 4.7. Variation of fractal dimension η versus p^* for steel with $m = 0$ and $m = 1$.

4.6 Concluding remarks

The results of this work show that Sih's concept on the strain energy density criterion can be applied in general to characterize metal damage at both the microscopic and macroscopic scale level. In particular, a relation is established to relate the microdamage and macrodamage parameters by application of the concept of fractals. The critical strain energy density functions described subcritical macrocrack growth and onset of macrocrack crack extension free of ambiguity.

References

[1] Sih, G.C., "Mechanics and physics of energy density theory", *Theoretical and Applied Fracture Mechanics*, Vol. 4, N3, pp. 157–173 (1985).
[2] Sih, G.C., "Thermomechanics of solids nonequilibrium and irreversibility", *Theoretical and Applied Fracture Mechanics*, Vol. 9, N3, pp. 175–198 (June 1988).
[3] Sih, G.C., "The strain energy density concept and criterion", *The Journal of the Aeronautical Soc. of India*, Vol. 37, NI, Part III, pp. 43–60 (February 1985).
[4] Ivanova, V.S., "Sinergeticheskaya model razrusheniya metallov i splavov po mekhanismu otryva tip 1") ("Synergistic model of metal and alloy failure by the break-off mechanism (type 1"). In Russian. *Fiz-Khim Mekhanika Materialov*, 22, 4, pp. 51–56 (1988).
[5] Ivanova, V.S., "Ustalostnoe razrushenie metallov" ("Fatigue Failure of Metals"). In Russian. – M.: *Metallurgiya*, pp. 258 (1963).
[6] Gladyshev, G.P., "Termodinamika i makrokinetika prirodnych ierarkhicheskikh processov" ("Thermodynamics and microkinetics of natural ierarchic processes") in Russian. – M.: *Nauka*, p. 287 (1988).

[7] Ivanova, V.S., Kunavin, S.A., Popov, E.A., and Poida, I.V., "K obosnovaniyu i raschetu postoyannoi razrusheniya dlya metallov i splavov" ("On interpretation and calculation of the failure constant for metals and alloys"). In Russian. *Metally*, N6, pp. 83–88 (1988).

[8] Naumov, I.I., Olkhovik, G.A., and Panin, V.E., "K obosnovaniyu termodinamicheskogo kriteriya prochnosti kristallov" ("On interpretation of the thermodynamic criterion of crystal strength"). In Russian. *Fiz-Khim. Mekhanika Materialov*, N3, V. 24, pp. 58–63 (1988).

[9] Fedorov, V.V., "Termodinamicheskie aspekty prochnosti i razrusheniya tverdyk tel" ("Thermodynamic aspects of strength and failure of solids"). In Russian. Tashkent, FAN of Uzbek SSR, p. 168 (1979).

[10] Ivanova, V.S., "K raschetu fiziko-khimicheskikh konstant metallov, svyazannykh s prochnoctyu mezhatomnoi svyazi" ("On calculation of physicochemical constants of metals related to interatomic bond strength"). In Russian. Khimiya metallicheskikh splavov.: M.: *Nauka*, pp. 196–204 (1983).

[11] Ivanova, V.S. and Terentyev, V.F., "Priroda ustalosti metallov" ("Nature of metal fatigue"). In Russian. *Metallurgiya*, p. 455 (1975).

[12] Ivanova, V.S., "Razrushenie metallov" ("Failure of metals"). In Russian. M.: *Metallurgiya*, p. 166 (1979).

[13] Haken, H., Advanced Synergetics, Springer, Berlin (1983).

[14] Ivanova, V.S. and Shanyavski, A.A., "Kolicchestvennaya fraktografiya. Ustalostnoe razrsuhenie" ("Quantitative Fractography. Fatigue Failure"). In Russian. *Metallurgiya*, Chelyabinsk Branch, p. 397 (1988).

[15] Mandelbrot, B.B., The Fractual Geometry of Nature, San Francisco, Freeman, 132 (1982).

[16] Richards, L.E. and Dempsey, B.D., Fractal Characterization of Fractal Surface in Ti4 A1–5.6 Mo–1.5 Cr. Scr. Met., V. 22, N. 5, pp. 687–689.

[17] Kleiser, T. and Bocek, M., "The fractal nature of slip in crystals" , *Z. Metall-kunde*, Bd. 77, N9, pp. 582–587 (1986).

Local and global instability in fracture mechanics

5.1　Introduction

The failure of structural elements may occur in different ways. The ultimate loading condition may be accompanied by plastic flow of the element, slow crack growth and/or fast crack propagation, depending on material strength, ductility and toughness as well as on structural size-scale. Initially uncracked elements of brittle material may reach their maximum loading capacity with a sudden crack formation and uncontrollable crack propagation, whereas cracked elements of ductile material may reach their ultimate condition with the progressive formation of a plastic hinge at the ligament. On the other hand, slow crack growth and plastic deformation can terminate with a rapid crack propagation.

Plastic Limit Analysis and Linear Elastic Fracture Mechanics reproduce the behaviour of very ductile and very brittle materials, respectively. For materials with intermediate properties, they are both unable to describe the behaviour under loading and the slow crack growth. Moreover, remarkable scale effects are present in fracture toughness testing. The mechanical specimen behaviour ranges from ductile to brittle only increasing the size-scale and keeping the geometrical shape unchanged. Large components fail due to rapid crack propagation, when the global behaviour is still linear elastic and the slow crack growth and plastic zones have not developed yet. On the contrary, small specimens fail in a ductile manner with a slow crack growth, so that the hardening and softening behaviours may be revealed up to the complete load relaxation and specimen separation.

In the present paper, the Strain Energy Density Theory of Sih [1, 2] is applied to analyze the integrity of structural members. Hardening or softening constitutive laws are used in the stress analysis, while the finite element method is applied consistently in both stress and failure analysis for each increment of loading.

G. C. Sih and E. E. Gdoutos (eds): Mechanics and Physics of Energy Density, 87–107.

As a consequence of the Strain Energy Density Theory, the rate of change of the strain energy density factor dS with respect to the crack growth rate da remained as a constant for each loading increment and specimen size-scale, i.e., dS/da = constant. The straight line relationship S versus a rotates in a counterclockwise direction around a common point as the loading increment is increased. The straight line translates as the size of geometrically similar specimens is increased.

Numerous examples are provided to illustrate how the Strain Energy Density Theory can be applied to predict different fail modes in metals (isotropic or kinematic hardening) and concretes (softening), ranging from plastic collapse to brittle fracture.

5.2 Strain energy density fracture criterion

In order to evaluate the crack growth increment at each loading step, the Strain Energy Density Theory will be applied, as proposed by Sih [1, 2]. It is based on the following fundamental assumptions.

- The stress field in the vicinity of the crack tip cannot be described in analytical terms because of the relative heterogeneity of the material. A minimum distance r_0 does exist below which it is meaningless to study the mechanical behaviour of the material from a "continuum-mechanics" point of view and to consider macroscopic crack growth increments.

- Outside such a core region of radius r_0, the strain energy density field can always be described by means of the following general relationship:

$$\left(\frac{\mathrm{d}W}{\mathrm{d}V}\right) = \frac{S}{r}, \tag{5.1}$$

   where the strain energy density factor S is generally a function of the three space coordinates.

- According to Beltrami's criterion, all the material elements in front of the crack tip, where the strain energy density is higher than the critical value $(\mathrm{d}W/\mathrm{d}V)_c^*$, fail (Figure 5.1).

When the following condition holds:

$$\Delta a = r^* = S_0 / (\mathrm{d}W/\mathrm{d}V)_c^*, \tag{5.2}$$

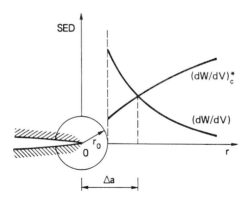

Figure 5.1. Crack growth increment Δa according to the strain energy density theory.

the crack may be considered as arrested, at least from a macroscopical point of view. On the other hand, when the crack growth increment is

$$\Delta a = r_c^* = S_c / (\mathrm{d}W/\mathrm{d}V)_c^*, \tag{5.3}$$

the unstable crack propagation takes place. S_c is a material constant and represents the strength of the material against rapid and uncontrollable crack propagation. S_c is connected with the critical value of the stress-intensity factor K_{1c} through the following equation (plane strain condition) [1]:

$$S_c = \frac{(1 + v)\ (1 - 2v)}{2\pi E}\ K_{1c}^2. \tag{5.4}$$

5.3 Strain-hardening materials

Material behaviour. The material behaviour is assumed to follow a multilinear hardening stress — strain relationship (Figure 5.2). The stress and strain increase up to the failure point c. Once the stress state surpasses the yield point e, permanent deformation or residual stress prevails: in fact, if the load is relaxed at point p, unloading occurs along the line pp'. The new constitutive relation then follows the path p'pc and irreversibility is thus considered.

For an undamaged material element, the critical value $(\mathrm{d}W/\mathrm{d}V)_c$ corresponding to failure is the area *o e c b* in Figure 5.2. At p, the material element is damaged, the area *o e p p'* being the dissipated strain energy density function, $(\mathrm{d}W/\mathrm{d}V)_d$. Therefore, the available energy density is:

$$\left(\frac{\mathrm{d}W}{\mathrm{d}V}\right)_c^* = \left(\frac{\mathrm{d}W}{\mathrm{d}V}\right)_c - \left(\frac{\mathrm{d}W}{\mathrm{d}V}\right)_d. \tag{5.5}$$

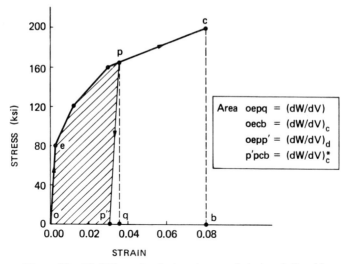

Figure 5.2. Multilinear hardening stress and strain relationship.

According to the incremental flow theory of plasticity [3], the center of the yield surface can move under plastic deformation and this motion is directed along the vector describing the current deviatoric stress state. The magnitude of the yield surface center rate is given by the scalar product of the stress rate vector and the unit outward normal vector. Moreover, the rate of the yield surface center is proportional to the coefficient β: *isotropic hardening* is achieved for $\beta = 0$ and *kinematic hardening* results for $\beta = 1$. Combined

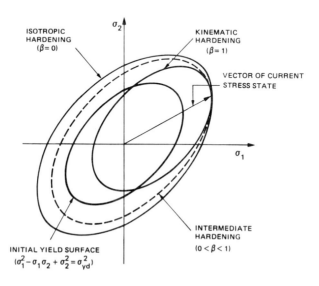

Figure 5.3. Comparison of subsequent yield surfaces for various hardening models.

hardening models can be described with values $0 < \beta < 1$. These possibilities are indicated in Figure 5.3 which shows the variations of the yield surface for different hardening models.

The failure of material elements ahead of the crack occurs when:

$$\left(\frac{dW}{dV}\right) = \left(\frac{dW}{dV}\right)^*_c.$$ (5.6)

For the material considered (Figure 5.2), this relationship is verified for $\varepsilon = 0.0475$ and $\sigma = 174$ ksi when $(dW/dV) = (dW/dV)^*_c = 6.55$ ksi.

Consider a three-point bending specimen (Figure 5.4), with width $b = 5$ in. length $l = 20$ in. and initial crack length $a_0 = 1$ in. in a condition of plane strain. The specimen material follows the stress-strain relationship given in Figure 5.2. Because of symmetry about the load line, only one-half of the specimen geometry needs to be considered. The finite element grid pattern is displayed in Figure 5.5. A total of 115 nodes and 18 elements are employed. Twelve nodes quadrilateral isoparametric elements and cubic displacement

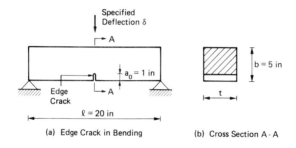

Figure 5.4. Three point bending specimen with an edge crack subjected to specified deflection δ.

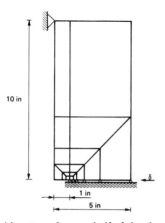

Figure 5.5. Finite element grid pattern for one-half of the three point bending specimen in Figure 5.4 (hardening material).

shape functions are used. The r^{-1} singularity in the strain energy density field is developed by means of the 1/9 to 4/9 nodal spacing of the elements adjacent to the crack tip.

Stress and energy state are computed by application of the Plastic Axisymmetric/Planar Structures (PAPST) computer program [4], which incorporates von Mises' yield criterion. The crack length increment and the strain energy density factor are computed by equating the strain energy density function dW/dV to the critical value $(dW/dV)^*_c = 6.55$ ksi (Figures 5.6 and 5.7). The abscissa of the intersection point P represents the crack length increment while the shaded area corresponds to the strain energy density factor.

As was previously stated, unstable crack growth is governed by S_c while local material damage depends on $(dW/dV)^*_c$. Energy dissipation and crack growth vary depending on the combination of material properties, loading step and specimen size [5–8]. The results obtained for the three-point bending specimen are useful to understand how the individual variables affect the behaviour of materials that follow a multilinear hardening stress-strain relationship (Figure 5.2).

Isotropic vs. kinematic hardening. Two different hardening models are considered. For Case A the rate vector of the yield surface center is zero, whereas for Case B it is maximum. Even though the process of energy

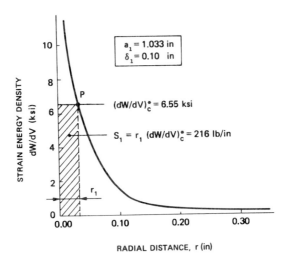

Figure 5.6. Variation of strain energy density function with radial distance for first increment of specified deflection ($\beta = 0$ and $\Delta\delta = 0.05$ in.).

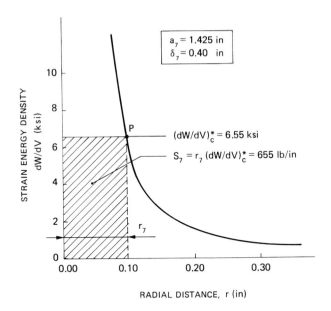

Figure 5.7. Variation of strain energy density function with radial distance for last increment of specified deflection ($\beta = 0$ and $\Delta\delta = 0.05$ in.).

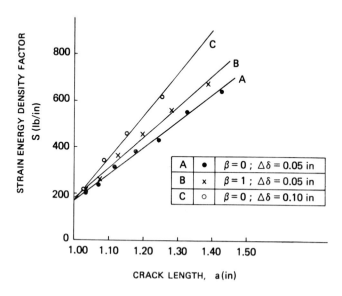

Figure 5.8. Strain energy density factor versus crack length for isotropic hardening ($\beta = 0$), kinematic hardening ($\beta = 1$) and different loading steps.

dissipation is a nonlinear phenomenon, the S versus a diagram is linear (Figure 5.8), i.e., the condition $dS/da = $ constant prevails. These straight line relations correspond to the situations when both crack growth and structure behaviour are stable. The curve of Case B has a larger slope than the curve of Case A: this means that, for a given material with $S_c = $ constant, a smaller subcritical crack length can be sustained in the case of kinematic hardening.

Effect of loading step. The effect of loading step is analyzed for a material with isotropic hardening. This effect can be clearly shown by plotting the S versus a diagrams (Figure 5.8). The two straight lines A and C correspond to $\Delta\delta = 0.05$ and 0.10 in. respectively. The slope dS/da increases as $\Delta\delta$ is increased: this implies that, for a given value of S_c, the subcritical crack length decreases if larger loading steps are chosen. Such a result is consistent with the experimental observations.

5.4 Strain-softening materials

Material behaviour. Damage of the material at the crack tip and crack growth increments will be computed on the basis of a uniaxial bilinear elastic-softening stress-strain relation (Figure 5.9). Stress may increase up to the ultimate strength (point U in Figure 5.9), while strain increases proportionally. Then, only strain may increase, while stress decreases linearly down to zero (point F in Figure 5.9). If the loading is relaxed when the representative point is in A, the unloading is assumed to occur along the line AO, so that the new bilinear constitutive relation is the line OAF. No permanent deformation is allowed by such a model, but only the degradation of the elastic modulus. In fact, the slope of line AO becomes lower when point A approaches point F. When point A is eventually coincident with F, the effective modulus E* vanishes, as well as the stress, and fracture or complete separation of the material occurs. On the other hand, when point A is on the elastic branch OU, unloading occurs along the same line and no degradation of the material takes place.

The present model simulates the mechanical damage by decreasing the elastic modulus, E, and the strain energy density which can be absorbed by a material element [6, 7, 9]. In fact, while for a non-damaged material element the critical value of the strain energy density, $(dW/dV)_c$, is equal to the area OUF (Figure 5.9), for a damaged material element with representative point in A, the decreased critical value of the strain energy density, $(dW/dV)_c^*$, is equal to the area OAF. In addition, as is shown in Figure 5.9, the area OUA represents the dissipated strain energy density, $(dW/dV)_d$, OAB the recoverable strain energy density, $(dW/dV)_r$, and BAF the additional strain energy density,

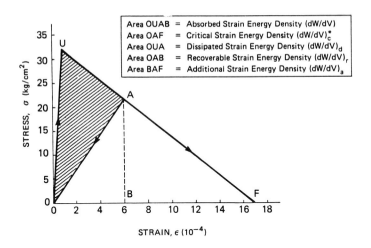

Figure 5.9. Bilinear elastic-softening stress and strain relationship.

$(dW/dV)_a$. In this way, the decreased critical strain energy density, $(dW/dV)_c^*$, can be expressed as follows:

$$\left(\frac{dW}{dV}\right)_c^* = \left(\frac{dW}{dV}\right)_c - \left(\frac{dW}{dV}\right)_d = \left(\frac{dW}{dV}\right)_r + \left(\frac{dW}{dV}\right)_a. \tag{5.7}$$

The above described model will be extended to the three-dimensional stress conditions, using the current value of the absorbed strain energy density, (dW/dV), as a measure of damage. In other words, the effective elastic modulus, E^*, and the decreased critical value of strain energy density, $(dW/dV)_c^*$, will be considered as functions of the absorbed strain energy density,

$$\left(\frac{dW}{dV}\right) = \left(\frac{dW}{dV}\right)_r + \left(\frac{dW}{dV}\right)_d.$$

Stress and strain in the softening condition A (Figure 5.9) can be expressed in terms of stress and strain in the ultimate and fracture conditions:

$$\sigma = E^*\varepsilon = \frac{\sigma_u \varepsilon_f}{(\varepsilon_f - \varepsilon_u) + \dfrac{\sigma_u}{E^*}}. \tag{5.8}$$

The three point bending specimen in Figure 5.4 is reconsidered for the above-described softening material with the following sizes: $b = t = 15$ cm,

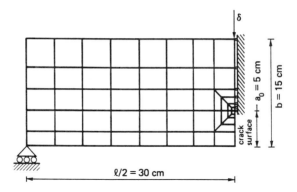

Figure 5.10.   Finite element grid pattern for one-half of the three point bending specimen in Figure 5.4 (softening material).

$l = 4b$, $a_0 = b/3$. The finite element mesh used for the specimen is represented in Figure 5.10. The Axisymmetric/Planar Elastic Structures (APES) finite element program [4] is applied at each loading increment. The idealization in Figure 5.10 utilizes 309 nodes and 52 elements and is considered in a condition of plane strain.

The damage-crack growth model is applied to the three point bending test specimen, considering a loading process for which the driving (input) parameter is the deflection δ (Figure 5.11). In this way, it is possible to follow even the softening stage of the load-displacement curve, as in a strain-controlled loading test. In Figure 5.11, the $P - \delta$ relationship is represented for

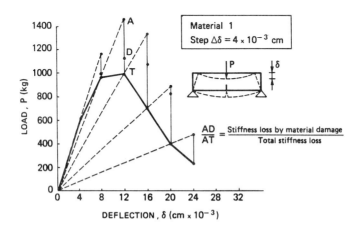

Figure 5.11   Load versus deflection relationship of three point bending specimen.

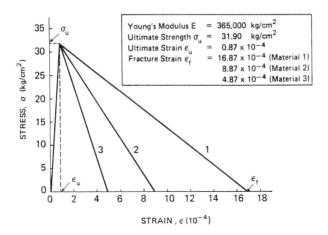

Young's Modulus E = 365,000 kg/cm²
Ultimate Strength σ_u = 31.90 kg/cm²
Ultimate Strain ε_u = 0.87 x 10⁻⁴
Fracture Strain ε_f = 16.87 x 10⁻⁴ (Material 1)
8.87 x 10⁻⁴ (Material 2)
4.87 x 10⁻⁴ (Material 3)

Figure 5.12.   Stress-strain curves for three different materials.

Material 1 (Figure 5.12) and for a constant deflection increment $\Delta\delta = 4 \times 10^{-3}$ cm. At the first step the stiffness is only 6.21% lower than the original one. Such a decrease is mostly due to the material damage at the crack tip. At the second step, the departure from linearity becomes significant, whereas at the third step the $P - \delta$ curve bends decidedly to the right and achieves its maximum. The segments AD and DT marked at the maximum (Figure 5.11) represent the decreases in the secant stiffness due to material damage and crack growth respectively. At the fourth step, the load P decreases and then the tangent stiffness results to be negative. This is due to the wide damage zone and to the extended crack, which make the specimen more and more flexible. At this stage, the contributions of material damage and crack growth to the secant stiffness decrease are almost the same. At the fifth and sixth steps, the load continues to decrease while the influence of crack growth increases sharply. When $\delta = 24 \times 10^{-3}$ cm, the load bearable by the specimen is only one fifth of the maximum, while the crack achieves the two thirds of the specimen width.

The distributions of the strain energy density, (dW/dV), and of the effective critical strain energy density, $(dW/dV)_c^*$, over the whole ligament are represented in Figure 5.13. While (dW/dV) decreases going away from the crack tip only inside each finite element, $(dW/dV)_c^*$ is a monotonic increasing function. The crack growth increment Δa is provided by the intersection of the two distributions. The positive jumps of the strain energy density function, (dW/dV), are due to the fact that the less damaged elements have a higher stiffness and, therefore, a higher load-bearing capacity. Such jumps represent a discrete softening effect ahead of the crack tip.

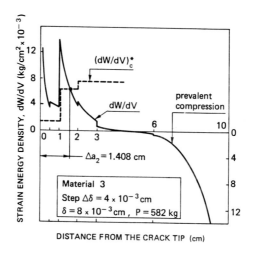

Figure 5.13. Distributions of the strain energy density function, (dW/dW), and of the effective critical strain energy density function, $(dW/dV)_c^*$, over the whole ligament.

Variation in the σ–ε *softening slope.* The numerical results will be discussed and their trends emphasized when varying the slope of the descending part of the curve σ–ε (Figure 5.12). The load-deflection responses for the deflection increment $\Delta\delta = 4 \times 10^{-3}$ cm are displayed in Figure 5.14. The shape of the curves is similar. The only substantial difference is that the area under them is approximately proportional to the area under the corresponding $\sigma - \varepsilon$ constitutive law.

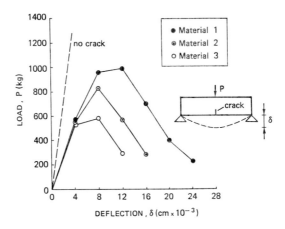

Figure 5.14. Load-deflection diagrams for three different materials (see Figure 5.12) and $\Delta\delta = 4 \times 10^{-3}$ cm.

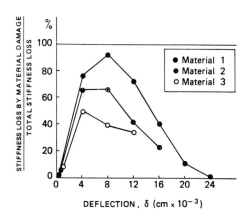

Figure 5.15. Ratio of stiffness loss by material damage to total stiffness loss as a function of the imposed deflection for three different materials (see Figure 5.12) and $\Delta\delta = 4 \times 10^{-3}$ cm.

In Figure 5.15, the ratio of stiffness loss by material damage to total stiffness loss (by material damage and crack growth) appears as a function of the imposed deflection δ. Such a ratio is equal to the ratio AD/AT in Figure 5.11. The general trend is that it tends to zero for very small as well as for very large deflections. For intermediate deflections, these curves present a maximum which increases by increasing the critical strain energy density of the material, $(dW/dV)_c$. In fact, for $\varepsilon_f \to \infty$, the material becomes elastic-perfectly plastic (without permanent deformation) and only material damage occurs, since $(dW/dV)_c$ and $(dW/dV)_c^*$ tend to infinity.

In Figure 5.16 the load P is reported against the crack growth, $(a - a_0)$. At the first steps, the load increases while the crack grows. Then, after reaching a maximum, the load decreases. The maximum represents the transition between stable and unstable structural behaviour. On the other hand, the transition between stable and unstable crack propagation depends on the achievement of the critical value of the strain energy density factor, S_c. Such a value has physical dimensions different from those of $(dW/dV)_c$ and, as will be shown later, this produces the scale effects recurrent in Fracture Mechanics. As a matter of fact, the crack instability may precede or follow the structural instability. This depends mainly on the structural scale. When the structure is relatively large, the crack instability precedes the traditional structural instability and Linear Elastic Fracture Mechanics can be correctly applied. On the other hand, when the structure is relatively small, as in the case of fracture toughness specimens, the structural collapse precedes the unstable crack propagation. The latter may occur only later, when the structural behaviour is in the softening stage (strain-controlled loading process) and the loading capacity is remarkably reduced. It is interesting to observe that, after

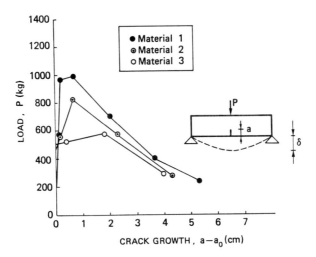

Figure 5.16. Load-crack growth diagrams for three different materials (see Figure 5.12) and $\Delta \delta = 4 \times 10^{-3}$ cm.

the first 2 cm of crack growth, the post-collapse structural behaviour is not considerably affected by the variation in $(dW/dV)_c$ (Figure 5.16).

In Figure 5.17, the values of the strain energy density factor, S, are reported against the crack growth, $(a - a_0)$. They appear nearly aligned. The results reported in Figure 5.17 relate only to small crack growth increments, when the structure is still in a stable $P - \delta$ ascending condition. Observe that the strain energy density factor rate, dS/da, has the same physical dimensions of the strain energy density and that it is always lower than the corresponding

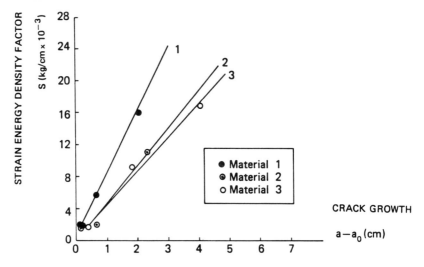

Figure 5.17. Strain energy density factor versus crack growth for three different materials (see Figure 5.12) and $\Delta \delta = 4 \times 10^{-3}$ cm.

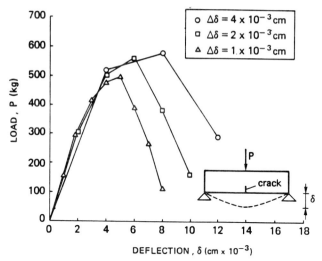

Figure 5.18. Load-deflection diagrams for Material 3 (see Figure 5.12) and three different loading increments.

$(dW/dV)_c$. The more important the damage is, compared with the stable crack growth, the smaller the ratio $(dS/da)/(dW/dV)_c$. We may expect this ratio to tend to zero when $\varepsilon_f \to \infty$ (elastic-perfectly plastic material) and to unity when $\varepsilon_f \to \varepsilon_u$ (elastic-perfectly brittle material).

Effect of loading step. In Figure 5.18, the load-deflection curves are displayed for material 3 and three different deflection increments: $\Delta\delta = 4 \times 10^{-3}$ cm, 2×10^{-3} cm, 1×10^{-3} cm. The maximum load, P_{max}, increases by increasing the deflection increment.

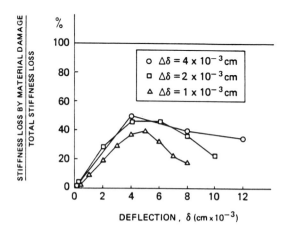

Figure 5.19. Ratio of stiffness loss by material damage to total stiffness loss as a function of the imposed deflection and for three different loading increments (Material 3 in Figure 5.12).

In Figure 5.19, the ratio of stiffness loss by material damage to total stiffness loss is reported as a function of the deflection, δ. It presents the same trends as discussed in the preceding section, i.e., damage prevails over crack growth only for intermediate crack lengths. Moreover, the stiffness loss by crack growth tends to prevail over the stiffness loss by material damage when the load step, $\Delta\delta$, decreases.

In Figure 5.20, the load P is plotted against the crack growth, $(a - a_o)$. As was already discussed in the preceding section the maximum of these curves represents the transition between stable and unstable structural behaviour and not necessarily between stable and unstable crack propagation. Also, the load-crack growth diagram is not very strongly affected by the load step, $\Delta\delta$, in the post-collapse condition.

In Figure 5.21, the values of the strain energy density factor, S, are reported versus the crack growth. As already shown in Figure 5.17, they appear aligned when both crack growth and structural behaviour are stable. The higher the deflection increment, $\Delta\delta$, is, the steeper the linear $S - a$ variation appears to be. This means that, the critical value of the strain energy density factor, S_c, being constant, with larger load steps the crack instability is achieved for smaller crack lengths.

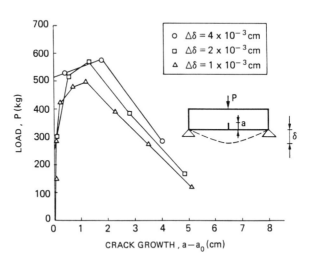

Figure 5.20. Load-crack growth diagrams for three different loading increments and Material 3 (see Figure 5.12).

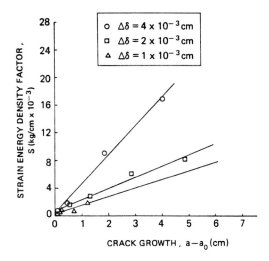

Figure 5.21. Strain energy density factor versus crack growth for three different loading increments and Material 3 (see Figure 5.12).

5.5 Size effects on strength and ductility

Because material damage and crack growth occur in a non self-similar fashion for each step of loading, specimens of different sizes appear to behave differently. This well-known size effect will be analyzed by the results obtained on the three-point bending specimen.

A dimensionless load parameter may be defined:

$$\frac{P}{(dW/dV)_c b^2} = \pi \left[\frac{\delta}{b}, \frac{E}{(dW/dV)_c}, \frac{\sigma_u}{(dW/dV)_c}, \nu, \frac{l}{b}, \frac{t}{b}, \frac{a_0}{b} \right], \tag{5.9}$$

as in the Buckingham's theorem for physical similitude and scale modelling. In equation (5.9) material toughness, $(dW/dV)_c$, and specimen width, b, have been used as the fundamental quantities. The dimensionless load parameter may be regarded as a function of the dimensionless deflection parameter δ/b only, if all others are kept constant.

In the same way, it is possible to define a dimensionless strain energy density factor:

$$\frac{S}{(dW/dV)_c b} = \Sigma \left[\frac{a}{b}, \frac{E}{(dW/dV)_c}, \frac{\sigma_u}{(dW/dV)_c}, \nu, \frac{l}{b}, \frac{t}{b}, \frac{a_0}{b} \right]. \tag{5.10}$$

Function Σ can be regarded as linear in a/b [5–8]:

$$\frac{S}{(dW/dV)_c b} = \frac{dS/da}{(dW/dV)_c}\frac{a - a_0}{b} + \frac{S_0}{(dW/dV)_c\ b} , \tag{5.11}$$

which may obviously be rearranged into the form [7]:

$$\tilde{S} = \tilde{S}'\ (\xi - \xi_0) + \tilde{S}_0. \tag{5.12}$$

The constants $\tilde{S}'$ and $\tilde{S}_0$ are dimensionless and scale independent. It follows that the slope of the $S - a$ diagram is constant varying the scale, i.e., $dS/da = $ constant, and that the intercept, S_0, is proportional to the scale b.

For a fixed $\Delta\delta/b$ ratio of 0.01 and isotropic hardening ($\beta = 0$), all the S versus $(a - a_0)$ diagrams (Figure 5.22) are parallel lines and the intercept S_0 is proportional to the scaling size b. Figure 5.22 shows the S versus $(a - a_0)$ shifting with b varying from 1 in. to 5 in. If, for example, the critical value of the strain energy density factor is $S_c = 150$ lb/in, then the limiting size is $b = 4.41$ in. For $b > 4.41$ in. only the unstable crack growth is possible.

Figure 5.23 relates to the softening Material 1 (Figure 5.12) and gives the straight line plots of S versus $(a - a_0)$ with b varying from 15 cm upwards. The critical crack growth decreases with increasing specimen size. With the critical value $S_c = 8 \times 10^{-3}$ kg/cm, the limiting size is $b = 120$ cm. Beyond this size, stable crack growth ceases to occur and failure corresponds to unstable crack propagation or catastrophic fracture.

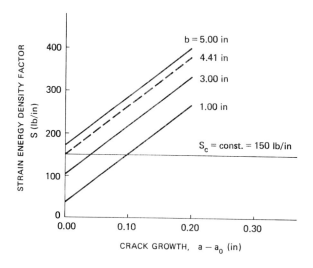

Figure 5.22.   Strain energy density factor versus crack growth by varying the specimen size-scale. Isotropic hardening material, $\Delta\delta/b = 0.01$.

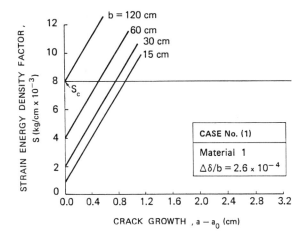

Figure 5.23. Strain energy density factor versus crack growth by varying the specimen size-scale. Softening material (No. 1 in Figure 5.12), $\Delta\delta/b = 2.6 \times 10^{-4}$.

The relations between $P/(\mathrm{d}W/\mathrm{d}V)_c\, b^2$ and δ/b are summarized in Figure 5.24. The vertical lines with arrows indicate the limiting values of δ/b as the critical strain energy density factor, $S_c = 8 \times 10^{-3}$ kg/cm, is reached. This corresponds to $K_{1c} = 144.35$ kg/cm$^{3/2}$ which is typical of concrete [10, 11]. It is evident that crack instability occurs for smaller dimensionless deflections of the specimen as size b increases.

It is now apparent that the quantity $S_c/(\mathrm{d}W/\mathrm{d}V)_c\, b$ must also enter into the dimensional analysis of equation (5.9). In fact, for estimating $P_{\max}$, it suffices to consider

$$\frac{P_{\max}}{(\mathrm{d}W/\mathrm{d}V)_c\, b^2} = \pi(S^*), \tag{5.13}$$

in which S^* is a dimensionless quantity:

$$S^* = \frac{S_c}{(\mathrm{d}W/\mathrm{d}V)_c\, b} = \frac{r_c}{b}. \tag{5.14}$$

Hence, all geometrically similar structures can be regarded as governed by S^*. This dimensionless quantity can be used to predict the load versus deflection behaviour for all specimen sizes [12].

In conclusion and recalling equation (5.12), it is possible to state that above the size (Figure 5.25):

$$b_{\max} = \frac{S_c}{(\mathrm{d}W/\mathrm{d}V)_c\, \tilde{S}_0} = \frac{r_c}{\tilde{S}_0}, \tag{5.15}$$

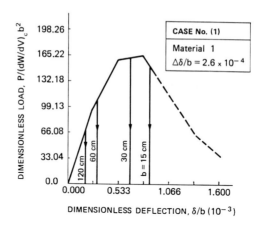

Figure 5.24. Dimensionless load-deflection diagram. Softening material (No. 1 in Figure 5.12), $\Delta\delta/b = 2.6 \times 10^{-4}$.

stable crack growth ceases to occur and the brittle failure is achieved when the $P - \delta$ curve is still in its linear initial course (Figure 5.24), whereas below the size (Figure 5.25):

$$b_{\min} = \frac{S_c}{(\mathrm{d}W/\mathrm{d}V)_c\,[\tilde{S}_0 + \tilde{S}'(1 - \xi_0)]} = \frac{r_c}{\tilde{S}_0 + \tilde{S}'(1 - \xi_0)}, \tag{5.16}$$

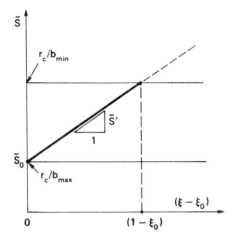

Figure 5.25. Dimensionless strain energy density factor versus crack growth.

unstable crack growth ceases to occur (even in the softening stage) and the progressive slow crack growth develops up to the complete specimen separation. On the other hand, for $b_{min} < b < b_{max}$, stable (or slow) crack growth is followed by unstable (or fast) crack propagation. Ductility appears therefore as a mechanical property depending on the size-scale rather than on the geometrical shape of the structure.

Acknowledgements

The author gratefully acknowledges the "Ministero Pubblica Istruzione" for the financial support received in the development of the present research. Thanks are also due to Prof. G.C. Sih, Director of the Institute of Fracture and Solid Mechanics, Lehigh University, Bethlehem Pa, U.S.A., for the very stimulating and fruitful discussions the author had with him during his sabbatical time at Lehigh University.

References

[1] Sih, G.C., "Some basic problems in fracture mechanics and new concepts", *Engineering Fracture Mechanics*, **5**, pp. 365–377, 1973.
[2] Sih, G.C. and Macdonald, B., "Fracture mechanics applied to engineering problems - strain energy density fracture criterion", *Engineering Fracture Mechanics*, **6**, pp. 361–386, 1974.
[3] Martin, J.B., *Plasticity: Fundamentals and General Results*, MIT Press, Cambridge, 1975.
[4] Gifford, L.N. and Hilton, P.D., "Preliminary documentation of PAPST nonlinear fracture and stress analysis by finite elements", David W. Taylor Naval Ship Research and Development Center, Bethesda MD, 1981.
[5] Sih, G.C. and Madenci, E., "Crack growth resistance characterized by the strain energy density function", *Engineering Fracture Mechanics*, **18**, pp. 1159–1171, 1983.
[6] Sih, G.C. and Matic, P., "A pseudo-linear analysis of yielding and crack growth: strain energy density criterion", *Defects, Fracture and Fatigue*, edited by G.C. Sih and J.W. Provan, Martinus Nijhoff Publishers, pp. 223–232, 1983.
[7] Carpinteri, A. and Sih, G.C., "Damage accumulation and crack growth in bilinear materials with softening", *I. Theoretical and Applied Fracture Mechanics*, **1**, pp. 145–159, 1984.
[8] Carpinteri Andrea, "Crack growth resistance in non-perfect plasticity: isotropic versus kinematic hardening", *I. Theoretical and Applied Fracture Mechanics*, **4**, pp. 117–122, 1985.
[9] Carpinteri, A., "Interpretation of the Griffith instability as a bifurcation of the global equilibrium", *Application of Fracture Mechanics to Cementitious Composites*, NATO Advanced Research Workshop, September 4–7, 1984, Northwestern University, edited by S.P. Shah, Martinus Nijhoff Publishers, pp. 287–316, 1985.
[10] Carpinteri, A., "Static and energetic fracture parameters for rocks and concretes", *Materials and Structures, RILEM*, **14**, pp. 151–162, 1981.
[11] Carpinteri, A., "Application of fracture mechanics to concrete structures", *Journal of the Structural Division, ASCE*, **108**, pp. 833–848, 1982.
[12] Carpinteri, A., "Notch sensitivity in fracture testing of aggregative materials", *Engineering Fracture Mechanics*, **16**, pp. 467–481, 1982.

A strain-rate dependent model for crack growth

6.1 Introduction

Classical continuum mechanics approach for determining the stress and/or strain distribution is solids in based on the assumption of a constitutive equation which is assumed to depend on the material while the interaction of material with local stress and/or strain is neglected. The form of the constitutive equation is preassumed and is taken to be the same for all material elements of the continuum. Such an approach, however, should be seriously questioned. When a nonhomogeneous stress and/or strain field is developed in a continuum, the various material elements experience different strain rates depending on their position and the load-time history and rate. The stress and strain response can vary from one location to another. Experiments on uniaxial tensile specimens for determining the mechanical response of materials show that the form of the stress-strain diagram of the material in tension changes with the rate and history of load application. The influence of strain rate and history is more profound on the post-yield behavior of the material.

A theory for the independent determination of the state of stress and strain in a continuum has been developed by Sih [1, 2]. It is based on the notion that irreversible material behavior can be uniquely described with the exchange of surface and volume energy density through the rate of change of volume with surface area. This is in contrast to the classical continuum mechanics approach that the rate change of volume with respect to surface is zero in the condition of equilibrium. Such a condition cannot be realized physically and has led to inaccurate predictions and inconsistencies. The strain energy density theory developed by Sih has been applied for the simultaneous determination of stress and failure to a host of problems of practical importance. Based on the concept of the plane of homogeneity in which the dissipated surface energy density is

G. C. Sih and E. E. Gdoutos (eds): Mechanics and Physics of Energy Density, 109–119.
© 1992 *Kluwer Academic Publishers. Printed in the Netherlands.*

equal at three mutually perpendicular directions, a one-to-one correspondence between uniaxial and multiaxial stress states can be achieved. The
results obtained by the strain energy theory deviate significantly from those of
the classical approach based on the separate analysis of stress and failure.

In the present paper, a model for a coupled stress and damage analysis is
developed by considering that the various material elements experience
different strain rates depending on their position and the rate of application of
the applied load. The method is applied to the case of a centre cracked panel
subjected to Mode-I loading. The results of the stress analysis are combined
with the strain energy density theory to determine the critical load for the
onset of crack growth.

6.2 Description of the method

The new methodology for stress analysis developed in the present work is
based on the notion that each material element of the continuum presents
different equivalent uniaxial stress-strain response depending on the load
time-history and rate. A host of uniaxial stress-strain curves corresponding to
different strain-rates is first considered. The continuum is discretized in smaller
elements and for the first increment of loading, all elements follow the same
equivalent uniaxial stress-strain curve corresponding to the rate of application
of the external loading. A stress analysis is then performed by using the
elastic-plastic finite element program PAPST [3]. This program uses twelve-
node isoparametric elements and a special singular crack tip element with two
side nodes placed at the 1/9 and 4/9 distance from the corner node at the crack
tip. The program is based on the J_2 flow theory of plasticity and the von Mises
yield criterion. The applied load is increased in a step-wise form and for each
increment of loading, each material element presents a different uniaxial
stress-strain behavior depending on the strain rate of the element. This strain
rate is defined as the difference between the equivalent strains at the beginning
and the end of the loading step divided by the time increment. More precisely,
the suggested methodology follows the steps below:

- For the first increment of loading, all material elements follow the same
  uniaxial stress-strain curve and a stress/strain analysis is performed.
- The principal strains ε_1, ε_2, ε_3 are determined at all Gaussian points of
  each element and an effective strain ε_{eff} is calculated from

$$\varepsilon_{\text{eff}} = \frac{1}{2}[(\varepsilon_1 - \varepsilon_2)^2 + (\varepsilon_2 - \varepsilon_3)^2 + (\varepsilon_3 - \varepsilon_1)^2]^{\frac{1}{2}}. \tag{6.1}$$

- The equivalent strain of each element is then defined as the average value
  of the effective strains at all Gaussian points of the element.
- The strain rate of each element is defined as the difference of the

equivalent strains between two successive loading increments divided by the corresponding time increment. Knowing the strain rate of the element, its equivalent uniaxial stress-strain response is determined from the data bank comprising of a host of uniaxial stress-strain curves corresponding to different strain rates.

● The preceding procedure is then repeated for each loading increment.

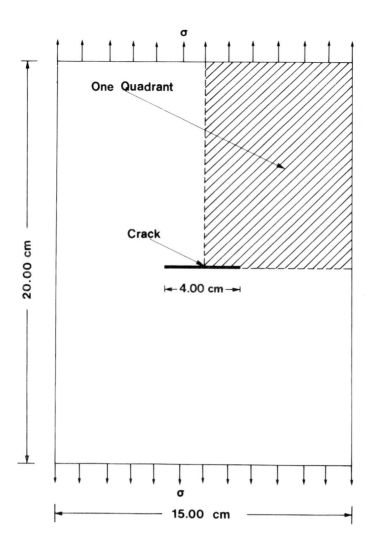

Figure 6.1. Geometry of the specimen.

6.3 Specimen geometry and material properties

The methodology developed in the preceding section for the determination of
the state of stress and strain in a continuum will be applied to the case of
a cracked plate subjected to a Mode-I loading as shown in Figure 6.1.
Considered is a rectangular panel of width $2b = 15$ cm and height $2h = 20$ cm
with a central through the thickness crack of length $2a = 4$ cm. The panel is
subjected to a monotonically increasing tensile stress σ in a direction
perpendicular to the crack plane. The thickness of the panel is considered very
large so that the state of stress can be approximated by plane strain.

The material of the panel is an aluminum alloy under the commercial
designation 1100–0 and has the elastic properties as shown in Table 6.1 in
which E is the modulus of elasticity, ν Poisson's ratio and σ_{ys} the yield stress.

Table 6.1. Mechanical properties of aluminum

Young's modulus E (MPa)	Poisson's ratio ν	Yield strength σ_{ys} (MPa)
3,310.6	0.33	44.8

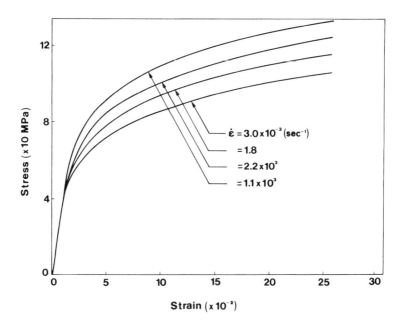

Figure 6.2. True stress-true strain curves of aluminum alloy 1100-0 corresponding to different strain rates.

The post yield true stress-true strain response of the material changes with the rate of application of strain. The curves of Figure 6.2 correspond to strain rates $\dot{\varepsilon} = 3.0 \times 10^{-3}$, 1.8, 2.2×10^2 and 1.1×10^3 (sec^{-1}) and were obtained from [4].

6.4 Stress analysis

Because of symmetry only one-quarter of the specimen geometry needs to be considered. Figure 6.3 presents the discretization of the one-quarter of the plate. Twenty-four elements with a total of 151 nodes were used. The first increment of loading corresponds to the applied stress that the most stressed node at the crack tip enters plastic deformation. The load is then increased at

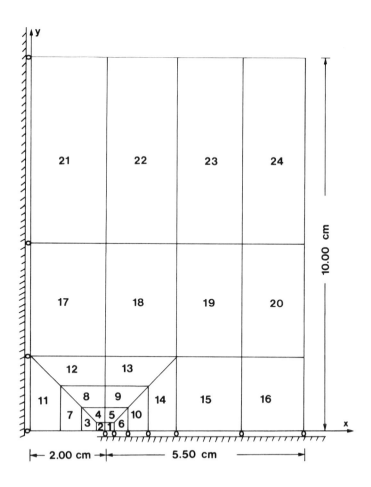

Figure 6.3. Finite element grid pattern for one quarter of the plate.

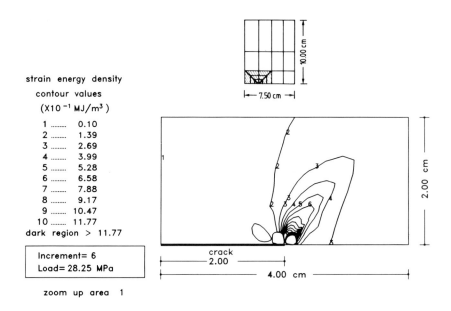

strain energy density
contour values
$(\times 10^{-1}$ MJ/m^3)

1	....	0.10
2	....	1.39
3	....	2.69
4	....	3.99
5	....	5.28
6	....	6.58
7	....	7.88
8	....	9.17
9	....	10.47
10	....	11.77

dark region > 11.77

Increment= 6
Load= 28.25 MPa

zoom up area   1

Figure 6.4.   Contours of the strain energy density function dW/dV for an applied stress $\sigma = 28.25$ MPa.

a step of 5 MPa occurring in a time interval of 1.875 sec. This corresponds to a constant applied strain rate at the plate boundary of $\dot{\varepsilon} = 3 \times 10^{-3}$ sec^{-1}.

Figure 6.4 presents the contour lines of the strain energy density function dW/dV defined by

$$\frac{dW}{dV} = \int_0^{\varepsilon_{ij}} \sigma_{ij} \, d\varepsilon_{ij} \qquad (6.2)$$

for the shaded area shown and for an applied stress $\sigma = 28.25$ MPa. The values of dW/dV corresponding to the various contours are shown in Figure 6.4. Note that dW/dV increases as the crack tip is approached.

As it was previously referred to, when a nonhomogeneous stress field is developed in a continuum, each material element has a different strain rate depending on the time-load history of the applied loading. Based on the results of the present study, the contour lines of strain rate in the neighborhood of the crack tip are plotted in Figure 6.5. Note that the strain rate increases as the distance from the crack tip decreases. Material elements closer to the crack tip present an equivalent uniaxial stress-strain curve corresponding to higher strain rates than material elements further away from the crack tip. Figure 6.6 presents the equivalent uniaxial stress-strain curve for material element number 1 (Figure 6.3). In the same figure, the stress-strain curve of the base material

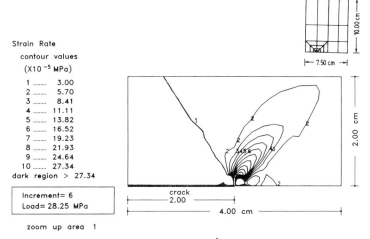

Strain Rate
contour values
(X10^{-5} MPa)

1	......	3.00
2	......	5.70
3	......	8.41
4	......	11.11
5	......	13.82
6	......	16.52
7	......	19.23
8	......	21.93
9	......	24.64
10	......	27.34

dark region > 27.34

Increment= 6
Load= 28.25 MPa

zoom up area 1

Figure 6.5. Contours of the equivalent strain rate $\dot{\varepsilon}$ for an applied stress $\sigma = 28.25$ MPa.

corresponding to an applied strain rate $\dot{\varepsilon} = 3 \times 10^{-3}$ sec^{-1} is also shown. Observe that the equivalent uniaxial stress-strain curve of the element is composed of different parts with increasing strain rates as the stress or strain increases. These strain rates are much higher than those assumed in plasticity. Values of strain rates $\dot{\varepsilon}$ at elements 1, 5 and 6 for ten loading increments are shown in Table 6.2.

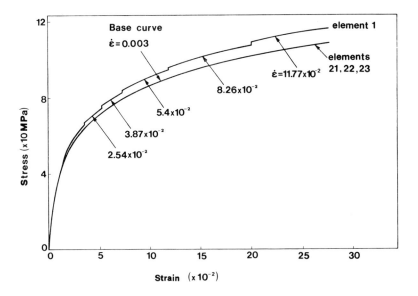

Figure 6.6. Equivalent uniaxial stress-strain response of material element 1 and of base material corresponding to an applied strain rate $\dot{\varepsilon} = 3 \times 10^{-3}$.

Table 6.2. Strain rates $\dot{\varepsilon}$ at elements 1, 5 and 6 for ten increments of applied stress σ.

Loading step	Applied stress	Strain rate ($\dot{\varepsilon}$) Element		
	(MPa)	1	5	6
1	3.25	0.00848	0.00514	0.00349
2	8.25	0.01332	0.00777	0.00530
3	13.25	0.01734	0.00799	0.00799
4	18.25	0.02546	0.00882	0.00600
5	23.25	0.03870	0.01323	0.00745
6	28.25	0.05406	0.02227	0.00841
7	33.25	0.08260	0.03625	0.00948
8	38.25	0.11778	0.04425	0.01601
9	43.25	0.23320	0.08700	0.03704
10	48.25	0.43960	0.17460	0.09020

The results obtained with the methodology presented in this work were compared with those based on the theory of plasticity when all material elements follow the same equivalent uniaxial stress-strain curves. Figures 6.7 and 6.8 display the variation of the normal and effective stresses σ_y and σ_{eff} with the distance r ahead of the crack for an applied stress $\sigma = 28.25$ MPa

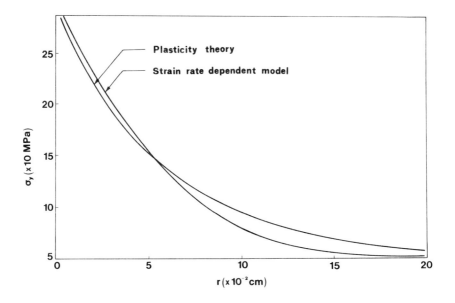

Figure 6.7. Variation of stress component σ^y versus distance r for an applied stress $\sigma = 28.25$ MPa calculated by plasticity theory and the method developed in this work.

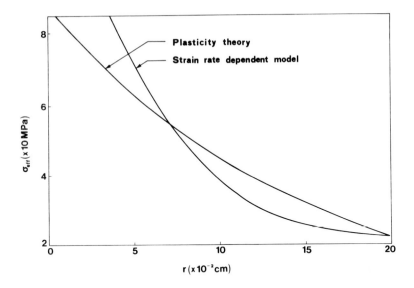

Figure 6.8. Variation of effective stress σ_{eff} versus distance r for an applied stress $\sigma = 28.25$ MPa calculated by plasticity theory and the method developed in this work.

calculated by the two theories. Results for the strain energy density function dW/dV are presented in Figure 6.9. Note that the present theory predicts higher values for the stresses σ_y and σ_{eff} as the crack tip is approached, while the strain energy density, dW/dV, is higher at all points along the crack ligament.

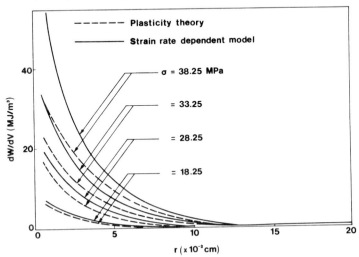

Figure 6.9. Variation of the strain energy density function dW/dV versus distance r for the values of the applied stress $\sigma = 18.25, 28.25, 33.25$ and 38.25 MPa calculated by plasticity theory and the method developed in this work.

6.5 Crack growth initiation

Having performed the analysis of the stress and energy fields of the cracked plate, study of the problem of initiation of crack growth needs a suitable failure criterion. As such, the strain energy density theory will be used. The fundamental quantity in this theory is the strain energy density function, dW/dV, which is given by equation (6.2). Fracture initiation takes place when the material element along the crack ligament and at a distance r_0 from the crack tip absorbs a critical amount of strain energy density, $(dW/dV)_c$. r_0 represents a ligament distance needed to define the resolution of the continuum mechanics analysis. It is the radius of the core region within which the continuum model ceases to be able to represent the state of stress and strain. For analysis purposes in the present study, r_0 was taken equal to 0.01 cm. The critical strain energy density function $(dW/dV)_c$ is calculated from the area underneath the true stress-true strain diagram of the material in tension and represents a material constant. It is the critical value of the energy required to fail a unit volume of material.

The critical applied stress for the onset of crack growth was calculated first when the analysis of the stress field was based on the theory of plasticity. For this reason, the finite element program PAPST was used. The critical strain energy density $(dW/dV)_c$ was calculated from the stress-strain curve of the base

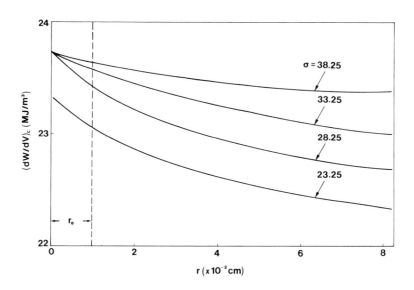

Figure 6.10. Variation of the critical value of the strain energy density function $(dW/dV)_c$ versus r for the values of the applied stress $\sigma = 23.25$, 28.25, 33.25 and 38.25 MPa.

material corresponding to a strain rate $\dot{\varepsilon} = 3 \times 10^{-3} \text{sec}^{-1}$. It was obtained that $(dW/dV)_c = 21.95 \text{ MJ/m}^3$. The critical value for the onset of crack growth was found to be $\sigma_i = 36.78 \text{ MPa}$.

The critical stress σ_i was also determined based on the new methodology presented in the present work. In this case, the various material elements ahead of the crack tip experience different strain rates depending on their position and the applied stress level and therefore follow different equivalent uniaxial stress-strain curves. Thus, the critical value of the strain energy density function $(dW/dV)_c$ defined as the area underneath the true stress-true strain curve of the material in tension changes with the position of the element along the crack ligament and the applied stress level. The variation of $(dW/dV)_c$ with r for an applied stress equal to $\sigma = 23.25, 28.25, 33.25$ and 38.25 MPa is plotted in Figure 6.10. Using the strain energy density theory, the critical stress for the onset of crack growth was found to be $\sigma_i = 31.4 \text{ MPa}$. Observe that this stress is about 15 percent smaller than the stress calculated using the theory of plasticity.

6.6 Concluding remarks

A strain rate dependent model for stress analysis of a solid was presented. It is based on the notion that when a nonhomogeneous stress field is developed in a solid, the various material elements experience different strain rates depending on their position and the rate of application of the external load. Therefore, each material element follows a different equivalent uniaxial stress-strain curve depending on the load time history. The method was applied to a centre-cracked rectangular panel subjected to a Mode-I loading. The results of the present study were compared with those obtained by plasticity theory. It was found that high strain rates are developed in the vicinity of the crack tip and the stresses and the strain energy density function are higher than those obtained by plasticity theory. The present study predicted a smaller value of the critical stress for the onset of crack growth than that obtained by plasticity theory. The predictions of this study are in agreement with those obtained by the strain energy density theory [1].

References

[1] Sih, G.C.,"Mechanics and physics of energy density theory", *Theoret. Appl. Fracture Mech.*, **4**, 157–173 (1985)
[2] Sih, G.C., "Thermomechanics of solids: Nonequilibrium and irreversibility", *Theoret. Appl. Fracture Mech.*, **9**, 175–198 (1988).
[3] PAPST, "Documentation of finite element program", Lehigh University, 1983.
[4] Lindholm, U.S. and Yeakley, L.M., "High strain-rate testing: Tension and compression", *Experimental Mechanics*, **8**, 1–9 (1968).

Extrusion of metal bars through different shape die: Damage evaluation by energy density theory

7.1 Introduction

The prediction of damage by yielding and/or fracture arising in mechanical forming and drawing entails considerable complications, not only in the assessment of stresses and strains but also in the selection of an appropriate failure criterion. Information on strain or energy localization induced in the material during its processing history can shed light on the desired combination of die shape with drawing rate so as to avoid or minimize the possibility of void or crack formation [1].

The flow of metal during extrusion [2] is a process that has been analyzed by many. What changes continuously with time is the flow pattern as the metal is pushed through the die. The material near the die is held back as the path of metal flow is narrowed down, whilst the central region can flow more easily through the die. A highly nonuniform change in the volume and shape of the local material elements can occur and lead to damage by crack initiation has been observed to occur in the center portion of the specimen. The location and size of damage depend on the die geometry and the rate of extrusion. Assessment of crack initiation in extrusion has lacked behind tests mainly because of the difficulties involved to supplement the plasticity theory with a suitable crack initiation criterion. This is not straightforward, because the procedure involves certain inherent shortcomings that should be recognized for otherwise misinterpretations could arise when comparing results with experiments. Keep in mind that the classical theory of plasticity or visco-plasticity is formulated by assuming that dilatational effects are small in comparison with distortional effects. Hence, the implementation of fracture criterion on stress solution obtained from plasticity theory that leaves out volume change may not be adequate because it has a major influence on fracture initiation.

G. C. Sih and E. E. Gdoutos (eds): Mechanics and Physics of Energy Density, 121–136.
© 1992 *Kluwer Academic Publishers. Printed in the Netherlands.*

Barring from the shortcomings in stress analysis just mentioned, an attempt is made in this work to characterize the metal extrusion process by focusing attention on the nonuniform change of the volume energy density function dW/dV in the local material elements. These nonuniformities are reflected by the changes in the element shape and volume which can be measured from changes in grid pattern attached to the specimens that undergo extrusion. The locations where distortion and dilatation would dominate in a nonlinear* elastic-plastic stress/strain field can be identified with the stationary values of the strain energy density functions [3–5]. On fundamental grounds, the metal extrusion process could be better described. Inclusion of the full dilatational effects into the stress analysis has been made [6]. This, however, would involve a major effort that is beyond the scope of this work.

7.2 Yielding/fracture initiation in plastic deformation

While attempts have been made to construct more complicated constitutive relations to supplement the classical theory of plasticity, analytical treatment of the metal forming process has not advanced simply because mechanical behavior of the metal could not be addressed without including damage. What is especially fundamental is the simultaneous description of yield and fracture. These two mechanisms occur in varying degrees at different locations; their proportion cannot be assigned arbitrarily but must be determined consistently in accordance with the load history of metal deformation.

Yield/fracture initiation. Because yield and fracture involve material damage at different scale level, any criterion that attempts to model both of these failure mechanisms would have to distinguish microdamage from macro-damage. As yielding is associated with permanent change caused by excessive distortion, fracture is related to volume change or excessive dilatation. These two events occur one after the other and never at the same location at a given *scale level* of observation [5]. That is, the location of macrodilatation may contain microdistortion but does not coincide with that of macrodistortion at the same instance and location. The two basic hypothesis [3, 4] of the strain energy density criterion that are relevant to predicting crack initiation in metal forming can be stated as follows:

---

* When plasticity is present, material behavior becomes nonlinear, the volume energy density function can no longer be divided into the distortional and dilatational component. The strain energy density theory [3–5] would determine the proportion of distortion and dilatation automatically without appealing to the principle of linear superposition.

- Failure initiation by yielding and fracture is assumed to coincide with locations of the maximum of $(dW/dV)_{max}$ and $(dW/dV)_{min}$ denoted, respectively, as $(dW/dV)_{max}^{max}$ and $(dW/dV)_{min}^{max}$.
- Failure by yielding and fracture is assumed to occur when $(dW/dV)_{max}^{max}$ and $(dW/dV)_{min}^{max}$ reach their respective critical values, say $(dW/dV)_y$ and $(dW/dV)_c$.

Uniqueness of failure location. It is because stationary values of dW/dV in a system may have many maxima and minima, the maximum among the many maxima and minima would first reach their threshold values. The pair $(dW/dV)_{max}^{max}$ and $(dW/dV)_{min}^{max}$ is unique. Analytical or numerical calculations can always be made to obtain the variations of dW/dV with the space variables as illustrated in Figure 7.1. It is guaranteed that there will be a maximum of dW/dV among the maxima and a corresponding maximum among the minima. Note that these are *relative* maximum and minimum with specific physical interpretations. The peak $(dW/dV)_{max}$ refers to yield because shape alteration

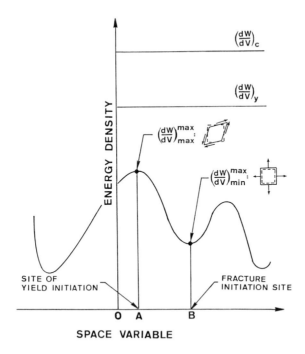

Figure 7.1. Stationary values of strain energy density function associated with yield and fracture initiation.

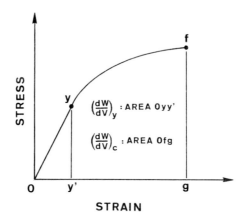

Figure 7.2. Critical values of volume energy density at yield and fracture.

associates with small volume change. This makes the quotation dW/dV a *relative maximum*. Volume change increases the denominator of dW/dV and makes dW/dV a *relative minimum* or $(dW/dV)_{min}$ which is associated with fracture.

Because $(dW/dV)_{max}^{max}$ is always greater than $(dW/dV)_{min}^{max}$ and $(dW/dV)_y <$ $< (dW/dV)_c$, yielding will always take place before fracture. The uniaxial stress and strain response in Figure 7.2 clearly shows that the area under the curve oyy' equivalent to $(dW/dV)_y$ must necessarily be smaller than ofg which is equal to $(dW/dV)_c$. If the thresholds $(dW/dV)_y$ and $(dW/dV)_c$ are displayed on the strain energy density diagram in Figure 7.1, it is obvious that as the load is increased the stress and strain would follow the curve from 0 to y and to f in Figure 7.2. Both $(dW/dV)_{max}^{max}$ and $(dW/dV)_{min}^{mx}$ would rise accordingly; $(dW/dV)_{max}^{max}$ would first meet $(dW/dV)_y$ at A followed by $(dW/dV)_{min}^{max}$ reaching $(dW/dV)_c$ at B. This shows that yielding will always precede fracture and they occur at different locations.

Successful application of the strain energy density criterion to the elastic-plastic behavior of metals in forming has been made [7]. The locations of yield and fracture initiation were determined for each time instance as the metal sheet is formed in steps.

7.3 Nonlinear behavior of extruded metal

In what follows, the nonlinear metal flow will be modelled. A steady state metal flow pattern is obtained experimentally. Deformed grid patterns in the meridian plane of the extruded billet are analyzed for four (4) different dies and two (2) different materials. The results are analyzed in conjunction with the strain energy density function to evaluate damage.

Die Ccnfigurations. To be considered are four die profiles; they are the conical, elliptical, hyperbolic and cosine shape. Referring to Figure 7.3, the parameters R_0 and R_i correspond, respectively, to the die radius at the entrance and exit and L represents the die length. The radius of the die profile at any distance z measured from the die entrance is R. The following four relations are used:

- *Conical Die*

$$R_i + (R_0 - R_i)[1 - (z/L)] \tag{7.1}$$

- *Cosine Die*

$$R = \frac{1}{2} [R_0 + R_i + (R_0 - R_i) \cos(\pi z/L)] \tag{7.2}$$

- *Hyperbolic Die*

$$R = [R_i^2 + (R_0^2 - R^2) (z/L)^2]^{\frac{1}{2}} \tag{7.3}$$

- *Elliptic Die*

$$R = [R_i^2 + (R_0^2 - R_i^2) (z/L)^2]^{\frac{1}{2}} \tag{7.4}$$

Mechanical and fracture properties. Two (2) different material types will be considered. The aluminum alloy is cold worked at room temperature but not annealed while the austenitic stainless steel is worked in a semi-hot state at 350°C. Their nonlinear stress and strain relations can be described by a two parameter equation of the form

$$\sigma_e = K(\varepsilon_e)^n \tag{7.5}$$

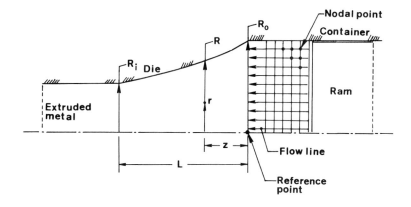

Figure 7.3. Extrusion of metal through die with flow lines.

Table 7.1.   Mechanical and fracture parameters for aluminum and stainless steel.

	Mechanical parameter		Critical available energy density
	K (J/cm^3)	n	$(\mathrm{d}W/\mathrm{d}V)_c$ (J/cm^3)
Aluminum	450	0.02	520
Stainless steel	1,260	0.22	1,250

in which σ_e is the effective stress

$$\sigma_e = \frac{1}{\sqrt{2}}\ \sqrt{(\sigma_1 - \sigma_2)^2 + (\sigma_2 - \sigma_3)^2 + (\sigma_3 - \sigma_1)^2} \tag{7.6}$$

and ε_e is the effective strain

$$\varepsilon_e = \frac{1}{\sqrt{2}}\ \sqrt{(\varepsilon_1 - \varepsilon_2)^2 + (\varepsilon_2 - \varepsilon_3)^2 + (\varepsilon_3 - \varepsilon_1)^2}\overset{\cdot}{} \tag{7.7}$$

In equations (7.6) and (7.7), σ_j and ε_j ($j = 1, 2, 3$) are the principal stresses and strains, respectively. As in plasticity, it is assumed that σ_e and ε_e coincide with the uniaxial tensile data. Given in Table 7.1 are the values of K and n for aluminum and stainless steel together with the corresponding values of the critical strain energy density function $(\mathrm{d}W/\mathrm{d}V)_c$ which is characteristic of the fracture toughness of the material. With the aid of equation (7.5), $(\mathrm{d}W/\mathrm{d}V)_c$ can be calculated from

$$\left(\frac{\mathrm{d}W}{\mathrm{d}V}\right)_c = \int_0^{\varepsilon_c} \sigma_e\ \mathrm{d}\varepsilon_e = \frac{K\varepsilon_c^{n+1}}{n+1} \tag{7.8}$$

in which ε_c is the ultimate strain. As mentioned earlier, $\mathrm{d}W/\mathrm{d}V$ can no longer be divided into a distortional and dilatational component because of nonlinearity. It, however, can be regarded as

$$\frac{\mathrm{d}W}{\mathrm{d}V} = \left(\frac{\mathrm{d}W}{\mathrm{d}V}\right)_p + \left(\frac{\mathrm{d}W}{\mathrm{d}V}\right)^* \tag{7.9}$$

where $(\mathrm{d}W/\mathrm{d}V)_p$ is the portion dissipated by plastic deformation. This corresponds to the area opp′ in Figure 7.4 while pqp′ represents the available energy density to do further damage as denoted by $(\mathrm{d}W/\mathrm{d}V)^*$ in equation (7.9). Therefore, either $(\mathrm{d}W/\mathrm{d}V)_c$ or $(\mathrm{d}W/\mathrm{d}V)^*_c$ can be used as threshold for fracture initiation.

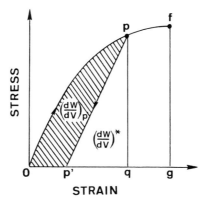

Figure 7.4. Response of material element illustrating energy dissipation.

Flow field characteristics. If the billet is long compared to its diameter, then the deformation pattern as shown in Figure 7.3 can be assumed to be quasi-static over most of the cycle. To be analyzed is the flow field in its steady-state condition so that time variations need not be considered. The velocity field can be determined from the changes in the lines of the grid pattern as the ram pushes the billet through the die. The flow function ψ is constant along each flow line. The initial values of ψ_j can be obtained as

$$\psi_j = \pi R_j^2 V_0 \tag{7.10}$$

where V_0 is the ram speed and R_j is the radius of a flow line before it experiences deformation. For axisymmetric extrusion, the flow function ψ depends on the coordinates r and z measured from, say the reference point in Figure 7.3.

Once ψ is known, the velocity components v_r and v_z follows immediately from

$$v_r = -\frac{1}{2\pi r}\frac{\partial\psi}{\partial z}, \qquad v_z = \frac{1}{2\pi r}\frac{\partial\psi}{\partial r}. \tag{7.11}$$

The strain rate components can thus be obtained as

$$\dot\varepsilon_r = \frac{\partial v_r}{\partial r}, \qquad \dot\varepsilon_\theta = \frac{v_r}{r}, \qquad \dot\varepsilon_z = \frac{\partial v_z}{\partial z},$$

$$\dot\gamma_{rz} = \frac{\partial v_r}{\partial z} + \frac{\partial v_z}{\partial r} \tag{7.12}$$

Numerical integration can then be performed along a flow line to obtain the effective strain rate

$$\dot{\varepsilon}_e = \sqrt{\frac{2}{3} \left[\dot{\varepsilon}_r^2 + \dot{\varepsilon}_\theta^2 + \dot{\varepsilon}_z^2 + \frac{1}{2} \dot{\gamma}_{rz}^2 \right]^{\frac{1}{2}}}. \tag{7.13}$$

If t_{jk} is the time elapsed for the kth point to be displaced along the jth flow line, then the corresponding effective strain can be obtained by numerical means as

$$(\varepsilon_e)_{jk} = \int_0^{t_{jk}} (\dot{\varepsilon}_e)_j \, dt. \tag{7.14}$$

The strain energy density function dW/dV or $\mathscr{W}$ can thus be calculated as

$$\frac{dW}{dV} \quad \text{or} \quad \mathscr{W} = \int_0^{t_{jk}} \dot{\mathscr{W}} \, dt \tag{7.15}$$

in which $\dot{\mathscr{W}}$ depends on $\dot{\varepsilon}_e$ and can be determined with the help of equation (7.5) relating σ_e and ε_e.

Numerical procedure. The experimental data are processed through a computer involving the steps outlined as follows:

- Grid pattern is constructed on the specimen involving vertical and horizontal lines as illustrated in Figure 7.3. Measurements are taken at the nodal points and used as input data until a steady flow pattern is established.

- A fourth order polynominal fit is used for ψ to calculate its value along each flow line by application of equation (7.10), i.e.,

$$\psi = c_1 r^2 + c_2 r^4 + c_3 r^6 + c_4 r^8, \quad \text{for } z = \text{const.} \tag{7.16}$$

- Smoothing of the flow function ψ data is made in both the r and z direction by assuming a uniform exit velocity at $z = L$ and applying the incompressibility condition.

- Once ψ is known, the velocity components v_r and v_z at the nodal points can be computed from equations (7.11) by using a three-point formula for numerical differentiation. Linear differentiation is then applied to obtain the velocity gradients $\partial v_r/\partial r$, $\partial v_r/\partial z$, etc. from which the strain rates in equations (7.12) can be obtained. The condition of incompressibility

$$\dot{\varepsilon}_r + \dot{\varepsilon}_\theta + \dot{\varepsilon}_z = 0 \tag{7.17}$$

was checked.

- The steady-state effective strain ε_e can be calculated numerically at the nodal points by using equation (7.13) and constant ε_e contours can be constructed graphically.
- By assuming that the stress and strain response is representable by equation (7.5), the strain energy density function at the nodal can be calculated as described earlier.

This completes the description of the experimental evaluation of dW/dV. There remains the determination of potential sites of failure initiation.

7.4 Analysis of failure initiation sites

Prior to application of the strain energy density criterion as discussed earlier, locations of the maximum of the minimum strain energy density functions need to be found. They would lie on the longitudinal axis because of axisymmetry.

Aluminum alloy. For the aluminum alloy, constant contours of ε_e are obtained for the four (4) die configurations specified in equations (7.1) to (7.4) inclusive. For the conical die, Figure 7.5 shows that the magnitude of ε_e increases with the distance z until the material passes beyond the exit section where ε_e regains its uniformity. A similar situation prevails for the cosine die in Figure 7.6 with a small difference in the distribution of ε_e at $z \simeq L$ as compared with the results

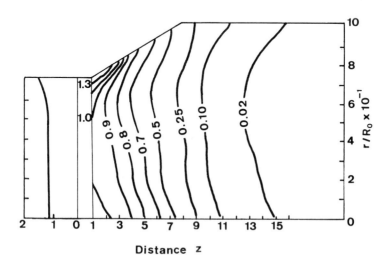

Figure 7.5. Contours of effective strain for aluminum through conical die.

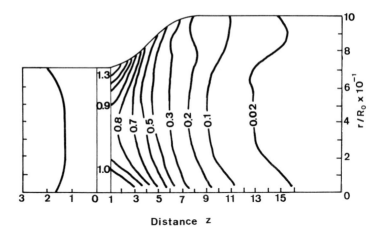

Figure 7.6.   Contours of effective strain for aluminum through cosine die.

for the conical die. Results for the hyperbolic die are given in Figure 7.7 where a more pronounced concentration of the σ_e contours occurred near the die shoulder. An even more severe concentration of σ_e is observed for an elliptic die as shown Figure 7.8 where high magnitude and gradient of σ_e appeared near the exit section. Making use of the results in Figures 7.5 to 7.8, $(dW/dV)_{min}$ are obtained for the four (4) die profiles; their values are shown in Table 7.2. The locations of all $(dW/dV)_{min}$ are near the exit section of the die.

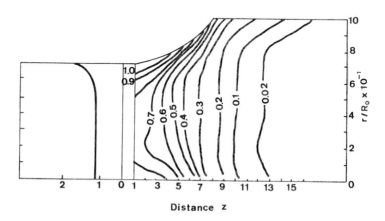

Figure 7.7.   Contours of effective strain for aluminum through hyperbolic die.

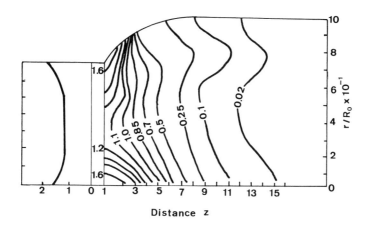

Figure 7.8. Contours of effective strain for aluminum through elliptic die.

Table 7.2. Maximum of the minimum in aluminum strain energy density function for different die profiles.

Die shape	Maximum/minimum of energy density $(dW/dV)_{\min}^{mx}$ (J/cm^3)
Conical	454
Cosine	517
Hyperbolic	504
Elliptic	531

Stainless steel. Similar results are found in the austenitic stainless steel. Contours of σ_e constant for the conical, hyperbolic and elliptic die are displayed in Figures 7.9, 7.10, and 7.11, respectively. Again, the effective strains increased as the material goes through the die. Die profile affected only the local disturbance of σ_e. The corresponding values of $(dW/dV)_{\min}^{mx}$ can be found in Table 7.3 and they also occurred near the exit.

Table 7.3. Maximum of the minimum strain energy density function in stainless steel for different die profiles.

Die shape	Maximum/minimum of energy density $(dW/dV)_{\min}^{mx}$ (J/cm^3)
Conical	1,159
Hyperbolic	1,033
Elliptic	907

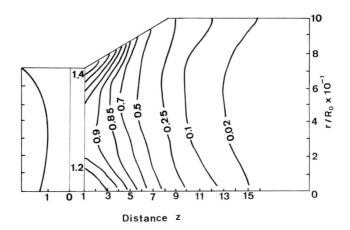

Figure 7.9. Contours of effective strain for stainless steel through conical die.

Deformed grid pattern. Displayed in Figures 7.12 to 7.15 are the deformed grid patterns for the extrusion of aluminum through the four (4) different die profiles. They all possess the general feature of increasing distortion and dilatation of the meshes as the material passes through the die. The die shape only affected the flow field next to the die boundary. Only in the case of the elliptic die that a crack initiated in the center portion of the billet near the exit. This confirmed with the quantitative assessment of $(\mathrm{d}W/\mathrm{d}V)_{\min}^{\max} = 531$ J/cm^3 which exceeded the $(\mathrm{d}W/\mathrm{d}V)_c = 520$ J/cm^3 in Table 7.1 for the aluminum. The $(\mathrm{d}W/\mathrm{d}V)_{\min}^{\max}$ of 512 J/cm^3 for the cosine die is only slightly less than critical. Crack initiation was not observed in the flow field of Figure 7.13.

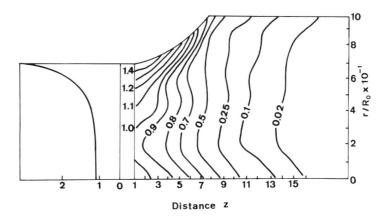

Figure 7.10. Contours of effective strain for stainless steel through hyperbolic die.

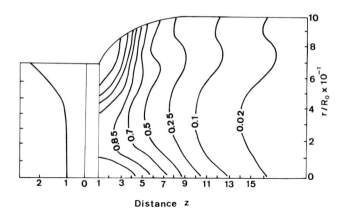

Figure 7.11. Contours of effective strain for stainless steel through elliptic die.

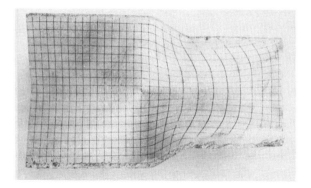

Figure 7.12. Flow pattern of aluminum through conical die.

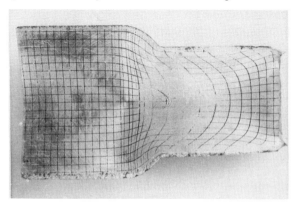

Figure 7.13. Flow pattern of aluminum through cosine die.

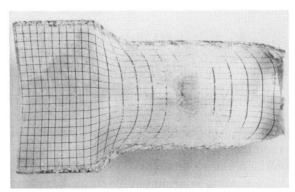

Figure 7.14. Flow pattern of aluminum through hyperbolic die.

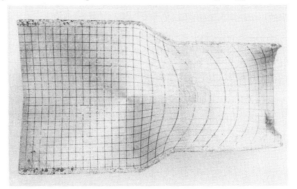

Figure 7.15. Flow pattern of aluminum through elliptic die.

Alterations in the flow pattern for the stainless steel material are exhibited in Figure 7.16 for the conical die; Figure 7.17 for the hyperbolic die; and Figure 7.18 for the elliptic die. The corresponding values of $(dW/dV)_{\min}^{\max}$ as given in Table 7.3 are not low when compared with the $(dW/dV)_c$ of 1,250 J/cm^3 in Table 7.1. Nevertheless, they were below threshold. This is consistent with the experimental observation of no crack initiation.

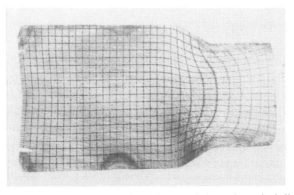

Figure 7.16. Flow pattern of stainless steel through conical die.

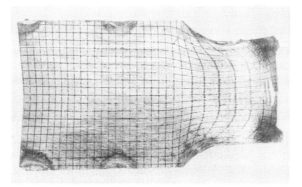

Figure 7.17. Flow pattern of stainless steel through hyperbolic die.

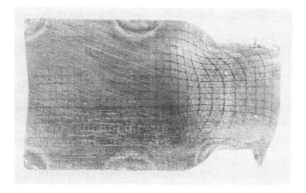

Figure 7.18. Flow pattern of stainless steel through elliptic die.

7.5 Conclusions

Because of the difficulties in obtaining realistic analytical predictions of the state of affairs in extrusion, an experimental approach was taken to obtain the strain field from which the volume energy density function was obtained numerically. Quantitative assessment on failure initiation agreed with experimental observation although the results showed small differences in the way with which die profiles affected the outcome. Influence on temperature changes is inherent in the experimental data although it could not be sorted out unless a more elaborate analytical analysis is performed. In other words, the determination of $(dW/dV)_p$ in equation (7.9) was not considered. A main objective of this work was to test the sensitivity of the experimental data for characterizing failure initiation by cracking as the material properties and die configurations are altered. The strain energy density criterion predicted the outcome successfully.

More refined prediction of metal extrusion behavior requires a theory that synchronizes the thermal and mechanical disturbances. The high rates of deformation in extrusion can cause substantial temperature rises. If these rises are sufficiently high, the metal can locally exceed its melting point giving rise to incipient melting. Even in the case of hot working, fir-tree like cracking could occur. This can be a particularly severe problem for the aluminum alloys. Past attempts in evaluating the transfer of heat during extrusion as an adiabatic process [8] is questionable because classical thermodynamics apply only to equilibrium phenomena where physical parameters are averaged over the time span of extrusion. The highly nonuniform deformation occurring in metal extrusion gives rise to nonhomogeneous and nonuniform thermal transients that are not desirable by equilibrium theories. Nonequilibrium theories of the type in [5, 6] must be used where the exchange of thermal and mechanical energy must be accounted for with or without phase changes of the material.

References

[1] Dragon, A., "A plasticity and ductile fracture damage study of void growth in metals", *J. of Engng. Fract. Mechs.,* Vol. 21, No. 4, pp. 875– (1985)

[2] Pearson, C.E., *The Extrusion of Metals,* Chapman and Hall, London (1953).

[3] Sih, G.C., *Introductory Chapters of Mechanics of Fracture,* Vol. I to VII, edited by G.C. Sih, Martinus Nijhoff Publishers, The Hague, (1972–1983).

[4] Sih, G.C. and Madenci, E., "Fracture initiation under gross yielding: strain energy density criterion", *J. of Engng. Fract. Mechs.,* Vol. 78, No. 3, pp. 667–677 (1983).

[5] Sih, G.C., "Thermal/mechanical interaction associated with micromechanics of material behavior", Institute of Fracture and Solid Mechanics, Lehigh University Technical Report, U.S. Library of Congress No. 87–080715, (February 1987).

[6] Sih, G.C., "Thermomechanics of solids: Nonequilibrium and irreversibility", *J. of Theoret. and Appl. Fract. Mechs.,* Vol. 9., No. 3, pp. 175–198 (1988)

[7] Sih, G.C., Chao, C.K., Hwu, Y.T. and Hsu, C.T., "Spring back in cold metal forming of A00–H steel: Non-homogeneous deformation", *J. of Theoret. and Appl. Fract. Mechs.,* Vol. 14, No. 2, pp. 81–99 (1990).

[8] Tanner, N. and Johnson, W., "Temperature distributions in some fast metal working operations", *Int. J. of Mech. Sci.,* Vol. 1, pp. 28–44 (1960).

Failure of a plate containing a partially bonded rigid fiber inclusion

8.1 Introduction

Fiber composite materials have gained popularity in engineering applications during the past few years due to their flexibility in obtaining the desired mechanical and physical properties in combination with light-weight components. For this reason, they are now being widely used for aerospace and other applications where high strengths and stiffness-to-weight ratios are required. These materials are usually made of glass or boron, graphite or other non-glass (for the so-called advanced composites) fibers embedded in an epoxy resin matrix. The way with which they are fabricated is very important for their efficient and reliable use in design. Quality control during the fabrication process is crucial and must be enforced.

Modelling the mechanical behavior of fiber composites is not a simple task. The materials are heterogeneous and are characterized by the presence of several types of inherent flaws. The basic failure modes are manifold including broken fibers, flaws in the matrix, debonded interfaces or their combination. The load transmission characteristics along the interface between the fibers and the matrix is crucial and it is not always possible to be accurately modelled.

Modelling of the failure behavior of fiber composites is made possible by using a number of simplifying assumptions. Among them the idealizations that the fibers are arranged parallel to themselves and no interaction effects between them take place and that the fibers are rigid, an assumption that is justified when the modulus of elasticity of the fiber is much higher than that of the matrix, are usually made. In this respect, the problem of a rigid fiber inclusion partially bonded to an infinite plate is of major importance.

G. C. Sih and E. E. Gdoutos (eds): Mechanics and Physics of Energy Density, 137–147.
© 1992 *Kluwer Academic Publishers. Printed in the Netherlands.*

The stress problem of the rigid fiber inclusion sealed-in an infinite plate was studied by Panasyuk et al [1]. Using the theory of complex potentials, they expressed the local stress and displacement fields around the ends of the inclusion in terms of two stress concentration factors. Chang [2] solved the problem of a two-dimensional rigid rectangular fiber inclusion in an infinite plate. His analysis was extended by Schneider and Conway [3] to nonrectangular fibers with particular emphasis given to the local geometry near the ends of the fiber. The problem of failure of a plate with a rigid fiber inclusion perfectly bonded to an infinite plate was studied by Gdoutos [4].

In the present paper, the failure behavior of a plate with a partially bonded rigid fiber inclusion is studied by using the results of stress analysis in conjunction with the strain energy density theory [5].

8.2 A partially bonded rigid elliptical inclusion in an infinite plate

Consider a rigid elliptical inclusion perfectly bonded to an elastic infinite plate except from a part A_S of its boundary forming an interfacial crack. The plate is characterized by its elastic constants κ and μ, where $\kappa = 3 - 4v$ and $\kappa = (3 - v)/(1 + v)$ for plane strain and plane stress, respectively, with μ being the shear modulus and v Poisson's ratio. A system of uniform stresses T and $;N$ is applied at infinity where the direction of T makes an angle φ with the x-axis.

The region outside the elliptical inclusion of the z plane is mapped to the infinite plate with a circular hole of the ζ-plane by the equation

$$z = m(\zeta) = R\left(\zeta + \frac{m}{\zeta}\right), \quad |\zeta| > 1, \tag{8.1}$$

where R $(R > 0)$ and m $(0 < m < 1)$ are real constants. For $m = 0$, the ellipse becomes a circle and for $m = 1$, a straight slit. Refer to Figures 8.1 (a) and 8.1 (b). The crack tips are located at points $\zeta_1 = \exp(i\theta_1)$ and $\zeta_2 = \exp(i\theta_2)$ and correspond to the points $z_1 = m(\zeta_1)$ and $z_2 = m(\zeta_2)$ of the z-plane. For $\zeta = e^\xi(\cos \eta + i \sin \eta)$ circles of radii $|\zeta| = e^{\xi_0}$ and straight lines $\eta = \eta_0$ in the ζ-plane define an orthogonal curvilinear coordinate system (ξ, η) in the z-plane. The stress components σ_ξ, σ_η and $\sigma_\xi{}^\eta$ referred to the system (ξ, η) are expressed by the following equations [6]

$$\sigma_\xi + \sigma_\eta = W(\zeta) + \overline{W(\zeta)}, \tag{8.2a}$$

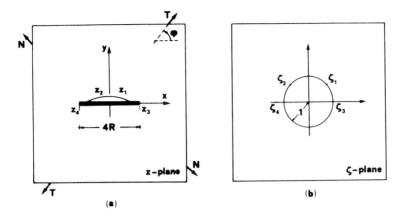

Figure 8.1. (a) A plate with a partially bonded rigid fiber inclusion and (b) its conformal mapping to the unit circle.

$$\sigma_\xi + i\sigma_{\xi\eta} = \frac{1}{2} W(\zeta) - \frac{m'(1/\overline{\zeta})}{2\zeta\overline{\zeta}m'(\zeta)} \overline{W}\left(\tfrac{1}{\zeta}\right)$$

$$+ \frac{1}{2}\left[1 - \frac{m'(1/\overline{\zeta})}{\overline{\zeta}\zeta m'(\zeta)}\right] \overline{W(\zeta)} \qquad (8.2b)$$

$$- \frac{\overline{\zeta}}{2\zeta}[m(\zeta) - m(1/\overline{\zeta})]\,\overline{W}'(\zeta),\,|\zeta| \geqslant 1,$$

where $W(\zeta)$ is a holomorphic function. For zero inclusion rotation $W(\zeta)$ is given by [6]

$$W(\zeta) = \frac{P(\zeta)\,X(\zeta)}{m'(\zeta)} \qquad (8.3)$$

where the functions $P(\zeta)$ and $X(\zeta)$ are expressed by

$$P(\zeta) = R\frac{N+T}{2}\left\{\zeta - D + \frac{1}{\kappa}\,\exp(i\theta_0 + \lambda\omega)\,(m + 2\delta\,e^{2i\varphi})\left(\frac{1}{\zeta^2} - \frac{d}{\zeta}\right)\right\} \quad (8.4a)$$

$$\qquad\qquad\qquad\qquad\qquad\qquad\qquad\qquad\qquad\qquad (8.4b)$$

$$X(\zeta) = (\zeta - \sigma_1)^{-\tau}\,(\zeta - \sigma_2)^{-\overline{\tau}}$$

with

$$D = \left(\cos\frac{\omega}{2} + 2\lambda\sin\frac{\omega}{2}\right)e^{i\theta_0}, \quad d = \left(\cos\frac{\omega}{2} - 2\lambda\sin\frac{\omega}{2}\right)e^{-i\theta_0} \quad (8.5a)$$

$$\delta = \frac{N-T}{N+T}, \qquad \omega = \theta_2 - \theta_1, \qquad \theta_0 = \frac{\theta_1 + \theta_2}{2}, \qquad \tau = \frac{1}{2} + \lambda i. \quad (8.5b)$$

$X(\zeta)$ is the Plemelj function. It is holomorphic in the ζ-plane cut along the arc Γ_D of the unit circle Γ. Only the branch of the function $X(\zeta)$ for which $\lim \; [\zeta X(\zeta)] = 1 \; \xi \to 0$ is considered.

The solution of the problem of a partially bonded rigid fiber inclusion in an infinite plate is obtained from the above solution as a limiting case by letting the minor axis of the ellipse to tend to zero ($m = 1$). As it is deduced from equations (8.2) to (8.5), the stress field becomes singular at the tips of the interfacial crack (the function $X(\zeta)$ becomes singular at those points) and at the ends of the inclusion which correspond to the roots $\zeta = \pm 1$ of the function $m'(\zeta)$. The singular behavior of the stress field at these points is studied in the sequel.

8.3 Local stress distribution and stress intensity factors

The local stress distribution around the crack tips or the fiber ends and the corresponding stress intensity factors will be determined.

Crack tip located arbitrarily along elliptical inclusion. Consider a local coordinate system $x_1 \, y_1$ at the right crack tip $z_1 = m(\zeta_1)$ so that the crack lies on the negative x_1-semi-axis as illustrated in Figure 8.2(a) ($m = 1$). For a point $t = \rho e^{i\alpha} \; (0 < \rho < R, 0 \leqslant \alpha \leqslant \pi)$ in the neighborhood of the crack tip, it can be put

$$\zeta \cong \zeta_1 + h_1 t, \quad h_1 = -\frac{i \; x p[(\theta_0 - \omega/2)i]}{R\sqrt{1 + m^2 - 2m \; \cos(2\theta_0 - \omega)}} \quad (8.6)$$

When the function $W(\zeta)$ given by equation (8.3) is referred to the system $x_1 \, y_1$ it takes the form

$$W(\zeta_1 + h_1 t) = \frac{K^{(1)}}{t^{\tau}} L^{\lambda i} + O(t^{\circ}), \quad (8.7)$$

where

$$K^{(1)} = \frac{N+T}{2}\sqrt{L}\,\overline{\tau} \, \exp\left(-\lambda\frac{\omega}{2} + i\frac{\omega}{2}\right) F^{(1)}, \quad (8.8a)$$

$$L = 2R \, \sin\frac{\omega}{2} \, \sqrt{1 + m^2 - 2m \; \cos(2\theta_0 - \omega)}, \quad (8.8b)$$

$$F^{(1)} = \frac{1 - e^{-(2\pi - \omega)\lambda} \, (m + 2\delta e^{2i\varphi}) \, e^{-(2\theta_0 - i\omega/2)}}{1 - m \, e^{-2i(\theta_0 - \omega/2)}}, \quad (8.8c)$$

In equation (8.7) the complex stress intensity factor $K^{(1)} = K_1^{(1)} - iK_2^{(1)}$ with dimensions stress (length)$^{\frac{1}{2}}$ has been introduced. From equations (8.2) and (8.7), the local stress field around the crack tip z_1 takes the form

$$\begin{bmatrix} \sigma_{x_1} \\ \sigma_{y_1} \\ \sigma_{x_1 y_1} \end{bmatrix} = \frac{e^{\lambda \alpha} K_1}{2\sqrt{\rho}} \begin{bmatrix} 3 \cos \hat{\alpha}_1 - \hat{e} \cos \hat{\beta}_1 - 2l_1 \sin \alpha \sin \hat{\gamma}_1 \\ \cos \hat{\alpha}_1 + \hat{e} \cos \hat{\beta}_1 + 2l_1 \sin \alpha \sin \hat{\gamma}_1 \\ \sin \hat{\alpha}_1 - \hat{e} \sin \hat{\beta}_1 + 2l_1 \sin \alpha \cos \hat{\gamma}_1 \end{bmatrix} \quad (8.9)$$

$$- \frac{e^{\lambda \alpha} K_2}{2\sqrt{\rho}} \begin{bmatrix} 3 \sin \hat{\alpha}_1 + \hat{e} \sin \hat{\beta}_1 + 2l_1 \sin \alpha \cos \hat{\gamma}_1 \\ \sin \hat{\alpha}_1 - \hat{e} \sin \hat{\beta}_1 - 2l_1 \sin \alpha \cos \hat{\gamma}_1 \\ - \cos \hat{\alpha}_1 - \hat{e} \cos \hat{\beta}_1 + 2l_1 \sin \alpha \cos \hat{\gamma}_1 \end{bmatrix}$$

such that

$$\hat{\alpha}_1 = 0.5 \alpha + \lambda \ln \frac{\rho}{L}, \qquad \hat{\beta}_1 = \frac{\alpha}{2} - \lambda \ln \frac{\rho}{L}, \qquad (8.10a)$$

$$\hat{\gamma} = 1.5 \alpha + \lambda \ln \frac{\rho}{L} - \tan^{-1}(2\lambda), \qquad l_1 = \sqrt{0.25 + \lambda^2},$$

$$\hat{e} = \exp[2\lambda(\pi - \alpha)] \qquad (8.10b)$$

Crack tip coincides with fiber end. This case is shown in Figure 8.2(b) and is obtained from the solution of the problem of elliptical inclusion with $m = 1$ and $\theta_1 = 0$ and $\zeta_1 = 0$. Consider a local coordinate system $x_1 y_1$ at the crack

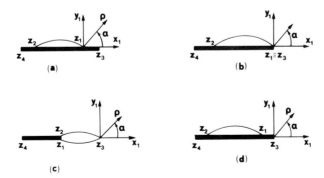

Figure 8.2. Different cases of a partially bonded rigid fiber inclusion in a plate.

tip z_1 which coincides with the fiber end $z_3 = m(1)$. For a point $t = \rho e^{i\alpha}$ $(0 < \rho << R, 0 \leqslant \alpha \leqslant \pi)$ in the neighborhood of $z_1 \equiv z_3$ it can be put

$$\zeta \cong 1 + \sqrt{\frac{t}{R}}. \tag{8.11}$$

Referring the function $W(\zeta)$ to the system $x_1 y_1$, it takes the form

$$W\left(1 + \sqrt{\frac{t}{R}}\right) = \frac{K^{(2)} L^{i\lambda/2}}{t^{1/2 + \tau/2}} + O\left(\frac{1}{t^{1/4}}\right), \tag{8.12}$$

with

$$K^{(2)} = \frac{(N + T)(4R)^{3/4}}{16\sqrt{\sin \omega/2}} \left[1 - D + (1 - d)(1 + 2\delta e^{2i\varphi}) \frac{e^{i\theta_0 - \lambda\omega}}{\kappa}\right], \tag{8.13a}$$

$$L = 4R \sin^2 \frac{\omega}{2}. \tag{8.13b}$$

In equation (8.12), the complex stress intensity factor $K^{(2)} = K_1^{(2)} - iK_2^{(2)}$ with dimensions stress (length)$^{3/4}$ has been introduced. From equations (8.2) and (8.12) the local stress field around the point $z_1 \equiv z_3$ takes the form

$$\begin{bmatrix} \sigma_{x_1} \\ \sigma_{y_1} \\ \sigma_{x_1 y_1} \end{bmatrix} = \frac{e^{\lambda\alpha} K_1}{2\rho^{0.75}} \begin{bmatrix} 3 \cos \hat{\alpha}_2 - \hat{e} \sin \hat{\beta}_2 - 2l_2 \sin \alpha \sin \hat{\gamma}_2 \\ \cos \hat{\alpha}_2 + \hat{e} \sin \hat{\beta}_2 + 2l_2 \sin \alpha \sin \hat{\gamma}_2 \\ \sin \hat{\alpha}_2 + \hat{e} \cos \hat{\beta}_2 + 2l_2 \sin \alpha \cos \hat{\gamma}_2 \end{bmatrix}$$
$$- \frac{e^{\lambda\alpha} K_2}{2\rho^{0.75}} \begin{bmatrix} 3 \sin \hat{\alpha}_2 - \hat{e} \cos \hat{\beta}_2 + 2l_2 \sin \alpha \cos \hat{\gamma}_2 \\ \sin \hat{\alpha}_2 + \hat{e} \cos \hat{\beta}_2 - 2l_2 \sin \alpha \cos \hat{\gamma}_2 \\ - \cos \hat{\alpha}_2 - \hat{e} \sin \hat{\beta}_2 + 2l_2 \sin \alpha \sin \hat{\gamma}_2 \end{bmatrix} \tag{8.14}$$

with

$$\hat{\alpha}_2 = 0.75\alpha + \frac{\lambda}{2}\ln\frac{\rho}{L}, \qquad \hat{\beta}_2 = 0.75\alpha - \frac{\lambda}{2}\ln\frac{\rho}{L} \tag{8.15a}$$

$$\hat{\gamma}_2 = \frac{7}{4}\alpha + \frac{\lambda}{2}\ln\frac{\rho}{L} - \tan^{-1}\left(\frac{2}{3}\lambda\right), \qquad l_2 = \sqrt{\frac{9}{16} + \left(\frac{\lambda}{2}\right)^2} \tag{8.15b}$$

Observe that the local stress field presents a (− 0.75) stress singularity.

Crack tip stress field. For $m = 1$ and $\theta_2 = -\theta_1 = \beta > 0$, the case in Figure 8.2(c) is obtained. When a local coordinate system is placed at point z_3, the function $W(\zeta)$ takes the form

$$W(\zeta) = \frac{K^{(3)}}{\sqrt{t}} + O(t^0) \tag{8.16}$$

with

$$K^{(3)} = K_1^{(3)} - iK_2^{(3)}$$

$$= \frac{(N + T)\sqrt{R}}{4} \left[\sin \frac{\beta}{2} - 2\lambda \cos \frac{\beta}{2} + \frac{e^{2\lambda\beta}}{\kappa} (1 + 2\delta\, e^{2i\varphi}) \right.$$

$$\left. \left(\sin \frac{\beta}{2} + 2\lambda \cos \frac{\beta}{2} \right) \right]. \tag{8.17}$$

Equation (8.17) gives the complex stress intensity factor $K^{(3)}$ at the crack tip z_3. The local stresses can be obtained by introducing $K^{(3)}$ into the well-known expressions of the singular stresses for a crack under mixed-mode conditions.

Fiber end stress field. Referring to Figure 8.2(d), the stress function $W(\zeta)$ referred to the fiber end takes the form

$$W(\zeta) = \frac{K^{(4)}}{\sqrt{t}} + O(t^0), \tag{8.18}$$

where

$$K^{(4)} = \frac{N + T}{4} \sqrt{R} \frac{\exp\left[\left(2\pi - \frac{\omega}{2} \right) \lambda - i \frac{\theta_0}{2} \right]}{\sqrt{\sin \left(\frac{\pi - \theta_0}{2} + \frac{\omega}{4} \right) \sin \left(\frac{\pi - \theta_0}{2} - \frac{\omega}{4} \right)}}$$

$$\times \left[\frac{\sin \left(\frac{\pi - \theta_0}{2} - \frac{\omega}{4} \right)}{\sin \left(\frac{\pi - \theta_0}{2} + \frac{\omega}{4} \right)} \right]^{\lambda i} F^{(4)} \tag{8.19}$$

and

$$F^{(4)} = 1 + D - (1 + d) \frac{e^{i\theta_0 + \lambda\omega}}{\kappa} (1 + 2\delta\, e^{2i\varphi}) \tag{8.20}$$

Equation (8.19) gives the complex stress concentration factor $K^{(4)}$ at the fiber end z_3. The local stresses can be obtained by introducing $K^{(4)}$ into the expressions of the singular stresses for a rigid fiber inclusion under mixed-mode conditions.

8.4 Failure initiation from the crack tip or the fiber end

Initiation of failure of a composite plate with a partially bonded rigid fiber inclusion takes place from the crack tips or the fiber ends since at these points high intensification of stresses takes place. For the determination of the critical failure stress and the fracture path, the strain energy density theory [5] is used. The fundamental quantity is the strain energy density function dW/dV which is expressed by

$$\frac{dW}{dV} = \frac{1}{4\mu}\left[\frac{\kappa + 1}{4}(\sigma_{x_1} + \sigma_{y_1})^2 - 2(\sigma_{x_1}\sigma_{y_1} - \tau_{x_1 y_1}^2)\right], \qquad (8.21)$$

where $\kappa = 3 - 4\theta$ or $\kappa = (3 - \theta)/(1 + \theta)$ for plane strain and generalized plane stress respectively, with θ being the Poisson ratio and μ the shear modulus.

The fracture path follows the direction at which the strain energy density function dW/dV at a distance ρ_0 from the failure site becomes minimum, while failure initiation occurs when dW/dV is equal to its critical value $(dW/dV)_c$. This value is a material parameter and it is equal to the area below the stress-strain diagram of the material in uniaxial tension.

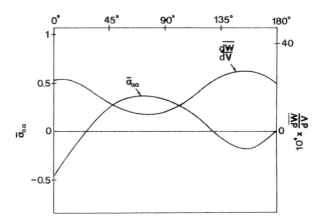

Figure 8.3. Variation of the normalized minimum strain energy density function $(d\bar{W}/dV)_{min} \times$ $\times 10^4$ and the normalized circumferential stress $\bar{\sigma}^{\alpha\alpha}$ versus the polar angle α for the case of Figure 8.2(a) with $\theta^0 = 90°$, $\omega = 90°$, $T = 0$, $\varphi = 0$, $\kappa = 1$ and $\rho = 10^{-5} R$.

Results for the critical stress for failure initiation and the initial fracture angle were obtained for different configurations of the fiber inclusion studied in Section 8.3. Figure 8.3 presents the variation of the normalized strain energy density function $d\bar{W}/dV = 16(\mu/N^2)(dW/dV)$ and the normalized circumferential stress $\bar{\sigma}_{\alpha\alpha} = 4(\rho/R)^{\frac{1}{2}}(\sigma_{\alpha\alpha}/N)$ versus the polar angle α for the case of Figure 8.2(a) when $\theta_0 = 90°$, $\omega = 90°$, $T = 0$, $\varphi = 0$, $\kappa = 1$ and $\rho = 10^{-5}R$. Observe that the strain energy density function becomes minimum at an angle $\alpha_c = 78.5°$ while the circumferential stress takes its maximum value at an angle $\alpha_c = 74.5°$. These values represent the critical fracture angles according to the minimum strain energy density and the maximum circumferential stress failure criteria. The results of the two theories are close to each other.

The variation of $(d\bar{W}/dV) \times 10^4$ and the fracture angle α_c versus the angular parameter β for the case of Figure 8.2(a) with $T = 0$, $\varphi = 0$, $\theta_0 = 90°$, $\omega = 2\beta$, $\kappa = 2$ and $\rho = 10^{-5}R$ is shown in Figure 8.4. Initiation of failure for this case when the crack is symmetrically located with respect to the fiber always takes place from the crack tip. Results for $(d\bar{W}/dV)$ and α_c versus the biaxiality coefficient $s = N/T$ when the crack extends along the whole length of the inclusion ($z_2 \equiv z_4$) and $\varphi = 0$, $\kappa = 2$ and $\rho = 10^{-5}R$ are presented in Figure 8.5. Finally, Figure 8.6 presents the variation of $(d\bar{W}/dV) \times 10^4$ versus the ratio l/L of the crack length l to $L = 4R$ for the case of Figure 8.2(d) when failure starts from the crack tip (point z_3) and the fiber end (point z_4). Observe that failure of the plate for small values of l/L starts from the crack tip while for higher values of l/L it starts from the fiber end.

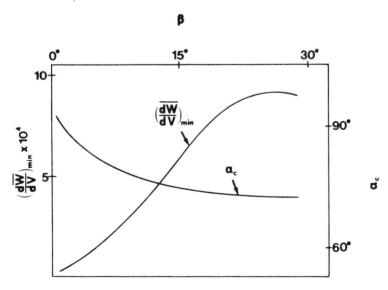

Figure 8.4. Variation of the normalized minimum strain energy density function $(d\bar{W}/dV)_{\min}$ $\times 10^4$ and the initial fracture angle a^c versus the angular parameter β for the case of Figure 8.2(a) with $\theta^0 = 90°$, $T = 0$, $\varphi = 0$, $\kappa = 2$ and $\rho = 10^{-5}R$.

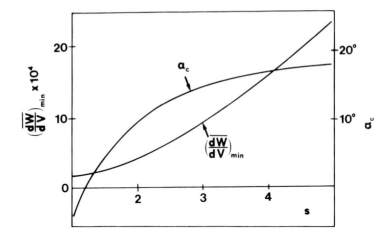

Figure 8.5. Variation of the normalized minimum strain energy density function $(\mathrm{d}\bar{W}/\mathrm{d}V)_{\min}$ $\times\ 10^4$ and the initial fracture angle a^c versus the biaxiality coefficient $s = N/T$ for the case of Figure 8.2(b) when $z^2 = z^4$ and $\varphi = 0$, $\kappa = 2$ and $\rho = 10^{-5}$ R.

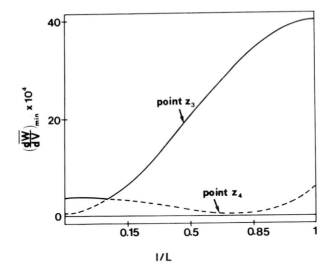

Figure 8.6. Variation of the normalized minimum strain energy density function $(\mathrm{d}\bar{W}/\mathrm{d}V)_{\min}$ $\times\ 10^4$ versus the ratio l/L for the case of Figure 8.2(c) when failure starts for the crack tip (point z^3) or the fiber end (point z^4).

References

[1] V.V. Panasyuk, L.T. Berezhnitskii and I.I. Trust, "Stress Distribution About Defects Such as Rigid Sharp-Angled Inclusions", *Problemy Prochn.* **7**, 3–9 (1972).
[2] C.S. Chang, "Some Two Dimensional Elastostatic Boundary Value Problems of Composite Materials", Doctoral Thesis, Report 818, Material Science Center, Cornell University.
[3] G.J. Schneider and H.D. Conway, "Effect of Fiber Geometry and Partial Debonding on Fiber-Matrix Bond Stresses", *J. Comp. Mat.* **3**, 116–135 (1969).
[4] E.E. Gdoutos, "Failure of a Composite with a Rigid Fiber Inclusion", *Acta Mechanica* **39**, 251–262 (1981).
[5] G.C. Sih, "Some Basic Problems in Fracture Mechanics and New Concepts", *J. Engng. Fract. Mech.* **5**, 365–377 (1973).
[6] E.E. Gdoutos and M.A. Kattis, "A Rigid Curvilinear Inclusion Partially Bonded in an Elastic Matrix", *Int. J. Engng. Sci.* **27**, 1287–1298 (1989).

Crack growth in rate sensitive solids

9.1 Introductory remarks

Crack growth in viscoelastic solids has been studied by the energy balance approach [1, 2] and the specification of critical parameters [3–6] associated with rupture in the vicinity of the crack tip. Previous works, however, involved complicated manipulations and expressions that were solvable only by numerical means. Considered in this treatise is the fracture criterion of energy density proposed by Sih [7]. The obtained results agreed well with experiments and circumvent complex mathematics.

9.2 Sih criterion

The criterion of Sih, though suggested initially to an elastic solid, was also applied for plastic [8, 9] and now for viscoelastic solids. It makes use of the energy density factor S which represents the intensity of the strain energy density function W. The critical value S_{cr} is assumed to be characteristic of the material while the direction of failure initiation is assumed to correspond with the maximum of minimum S such that the condition

$$\text{Min } S(\theta) = S_{cr} \tag{9.1}$$

corresponds to the onset of crack initiation. If r denotes the distance from the location of crack initiation, then

$$S = \lim_{r \to 0} rW. \tag{9.2}$$

G. C. Sih and E. E. Gdoutos (eds): Mechanics and Physics of Energy Density, 149–154.
© 1992 Kluwer Academic Publishers. Printed in the Netherlands.

For nonelastic solids as elastoplastic or viscoelastic W should be referred to as the free energy function.

9.3 Linear viscoelastic solid

Prior to the application of the Sih criterion to viscoelastic solids, the form of W of linear viscoelastic solids is required. Let σ_{ij} be written as

$$\sigma_i{}^j = \int_0^t C_{ijkl}(t-\tau)\, d\varepsilon_{kl} \equiv \bar{C}_{ijkl}[\varepsilon_{kl}], \tag{9.3}$$

in which $C_{ijkl}(t)$ stands for the kernel and $\bar{C}_{ijkl}$ shall be referred to as the operator in creep.

If ε_{ij} are limited to infinitesimal strains, then it can be divided into an elastic component ε^e_{ij} and a viscous component ε^v_{ij} as follows:

$$\varepsilon_{ij} = \varepsilon^e_{ij} + \varepsilon^v_{ij} = \int_0^t K_{ijkl}(t-\tau)\, d\sigma_{kl} \equiv \bar{K}_{ijkl}[\sigma_{kl}], \tag{9.4}$$

such that

$$\varepsilon^e_{ij} = K^0_{ijkl}\,\sigma_{kl}, \qquad \varepsilon^v_{ij} = \int_0^t \dot{K}_{ijkl}(t-\tau)\sigma_{kl}(\tau)\, d\tau. \tag{9.5}$$

The compliance operator $\bar{K}_{ijkl}$ in equation (9.4) is the inverse of $\bar{C}_{ijkl}$ and $K_{ijkl}(t)$ is the corresponding kernel of the compliance with $K^0_{ijkl} = K_{ijkl}(0)$. For a linear constitutive relation, W is a quadratic form of the strains:

$$W = \frac{1}{2}C^0_{ijkl}\,\varepsilon^e_{ij}\,\varepsilon^e_{kl} + C^1_{ijkl}\,\varepsilon^e_{ij}\,\varepsilon^v_{kl} + \frac{1}{2}C^2_{ijkl}\,\varepsilon^v_{ij}\,\varepsilon^v_{kl}. \tag{9.6}$$

The tensor C^0_{ijkl} corresponds to the instantaneous moduli $C_{ijkl}(0)$ while C^1_{ijkl} and C^2_{ijkl} can be expressed in terms of C^0_{ijkl} and the long term moduli $C_{ijkl}(\infty)$. Additional simplifications can be made if the linear viscoelastic solid is isotropic. Let the isotropic deviatoric stresses s_{ij} and strains e_{ij} be defined as

$$s_{ij} = \sigma_{ij} - p\delta_{ij}, \qquad e_{ij} = \varepsilon_{ij} - e\delta_{ij}, \tag{9.7}$$

in which δ_{ij} are the components of the Kronecker delta. The quantities e_{ij} and s_{ij} are related as

$$e_{ij} = \frac{1}{2\mu_0} s_{ij} + \int_0^t \dot{K}(t-\tau) s_{ij}(\tau) \, d\tau. \tag{9.8}$$

In equations (9.7), p and e stand for the isotropic stress and strain, respectively,

$$p = \frac{1}{3}\sigma_{kk}, \qquad e = \frac{1}{3}\varepsilon_{kk} \tag{9.9}$$

such that for elastically compressible solids

$$p = \kappa e, \qquad \kappa = \frac{2\mu_0(1+v_0)}{1-2v_0} \tag{9.10}$$

with κ being the elastic bulk modulus and μ_0 and v_0 are, respectively, the shear modulus and Poisson's ratio. It can be shown that for this type solids

$$W = \frac{3}{2}\kappa e^2 + \mu_0 \, e^e_{ij} \, e^e_{ij} + \mu_v \, e^v_{ij} \, e^v_{ij}, \tag{9.11}$$

where μ_v is given by

$$\mu_v = \frac{1}{K(\infty) - K(0)}. \tag{9.12}$$

Once the specific form of W is known, the criterion of Sih can be readily applied to determine the stationary values of W for obtaining the locations of failure initiation.

9.4 Crack growth in uniformly applied stress field

Consider the geometry of a center crack of length $2l_0$ where a uniform tensile of magnitude σ is applied at infinity and maintained constant thereafter. A coordinate system (x_1, x_2) is used such that the origin is placed at the middle of the crack with the x_2-axis normal to it. Because of symmetry, the crack would propagate along the x_1-axis. Application of the correspondence principle yields the viscoelastic solution for the stresses along the x_1-axis or $\theta = 0$:

$$s_{11} = s_{22} = \left(\frac{1-2\hat{v}}{3}\right) \frac{K_1}{\sqrt{2\pi r}},$$

$$s_{33} = -\frac{2(1 - 2\hat{v})}{3} \frac{K_1}{\sqrt{2\pi r}}. \tag{9.13}$$

In equation (9.12), K_1 is the Mode I stress intensity factor

$$K_1 = \sigma\sqrt{\pi l} \tag{9.14}$$

and $\hat{v}$ is the Poisson ratio operator

$$\hat{v}[f] = v_0 f(t) + \int_0^t L(t - \tau) f(\tau)\, d\tau \tag{9.15}$$

in which the kernel $L(t)$ is a linear function of $K(t)$. Making use of equations (9.8), (9.11) and (9.13), it is found that

$$W = \frac{\sigma^2}{2\mu^0}\left\{\frac{1 - 2v_0}{2}\frac{l}{x_1 - l} + \frac{1 + v^*}{1 + v_0}\left[\int_0^t \dot{L}(t - \tau)\sqrt{\frac{l(\tau)}{x_1 - l(\tau)}}\,d\tau\right]^2\right\}, \tag{9.16}$$

where

$$v^* = 1 + \frac{3\mu_v}{2\mu_0(1 + v_0)}. \tag{9.17}$$

The integral in equation (9.16) is bounded and its corresponding intensity can be found as

$$S = \frac{1 - 2v_0}{4\mu_0}\sigma^2 l. \tag{9.18}$$

The condition for overloading the crack is $S > S_{cr}$ at $t = 0$ with $l = l_0$ and the crack would grow. For $S < S_{cr}$ at $t = 0$, the crack would be stationary such that

$$S = \frac{\sigma^2 l}{2\mu_0}\left[\frac{1 - 2v_0}{2} + \frac{1 + v^*}{1 + v_0}L^2(t)\right]. \tag{9.19}$$

In general, the applied stress σ would be bounded as $\sigma_* < \sigma < \sigma^*$, where $\sigma < \sigma_*$ corresponds to no crack growth while $\sigma > \sigma^*$ to crack growth. These stresses are given by

$$\sigma^* = 2\left[\frac{\mu_0}{1 - 2v_0}\frac{S_{cr}}{l}\right]^{1/2} \tag{9.20}$$

and

$$\sigma_* = \left\{ \frac{8\mu_0(1+v_0)}{3(1-2v_0)} \frac{S_{cr}}{l} \left[1 + \frac{\mu_v(1-2v_0)}{2\mu_0(1+v_0)}\right]^{-1} \right\}^{\frac{1}{2}} \tag{9.21}$$

The time to initiate fracture, say t_d, can be determined from equation (9.19) by letting $S = S_{cr}$. In contrast to the criteria in [10, 11] used for analyzing crack growth in viscoelastic solids [2, 4], the present work yields a nontrivial σ_* for the Maxwell solid with $\mu_v = 0$:

$$\sigma_* = 2 \left[\frac{2\mu_0(1+v_o)}{3(1-2v_0)} \frac{S_{cr}}{l} \right]^{\frac{1}{2}}. \tag{9.22}$$

The above results can be used to compare with experimental data and determine the predictive capability of the various fracture criteria.

9.5 Conclusions

The strain energy density criterion of Sih [7] has been applied to determine the crack growth characteristics in a linear viscoelastic solid. In contrast to previous works on this subject which utilized energy balance and specific surface energy, application of the strain energy density criterion is straightforward once the energy density function is referred as the free energy and as such is known for a specific material. Crack growth behavior in elastic-plastic solids can be found in [8, 9].

References

[1] Williams, M.L., "Stress singularities, adhesion and fracture", *Proceedings of Fifth U.S. National Congress of Applied Mechanics*, pp. 451–464 (1966).
[2] Nikitin, L.V., "Application of the Griffith's approach to analysis of rupture in viscoelastic bodies", *Int. J. Fracture*, **24**, 2, pp. 149–157 (1984).
[3] Wnuk, M.P. and Knauss, W.G., *Delayed fracture in viscoelastic-plastic solids*, California Institute of Technology Report, prepared for NASA, GALCIT SM 68–8 (1968).
[4] Kostrov, B.V. and Nikitin, L.V., "Some general problems of mechanics of brittle fracture", *Arch. Mech. Stosaw.*, **22**, 6, pp. 749–776 (1970).
[5] McCartney, L.N., "Crack growth in viscoelastic materials", *Int. J. Fracture*, **13**, pp. 641–654 (1973).
[6] Schapery, R.A., "A theory of crack initiation and growth in viscoelastic media", *Int. J. Fracture*, **2**, 1, pp. 141–159 (1975).
[7] Sih, G.C., "Some basic problems in fracture mechanics and new concepts", *Engng. Fracture Mech.*, **5**, 2, pp. 365–377 (1973).

[8] Sih, G.C. and Chen, C., "Non-self-similar crack growth in an elastic-plastic finite thickness plate", *J. of Theoretical and Applied Fracture Mechanics,* 3, 2, pp. 125–139 (1985).

[9] Sih, G.C. and Tzou, D.Y., "Plastic deformation and crack growth behavior", in *Plasticity and Failure Behavior of Solids*, Series on Fatigue and Fracture, Vol. III, edited by G.C. Sih, A.J. Ishlinsky and S.T. Mileiko, Kluwer Academic Publishers, The Netherlands (1990).

[10] Griffith, A.A., "The phenomenon of rupture and flow in solids", *Phil. Trans. Roy. Soc.,* **A221**, pp. 163–198, (1920).

[11] Panasyuk, V.V., *Limited equilibrium of the brittle solids with cracks*, Kiev, Naukova Dumka, p. 246 (in Russian) (1968).

Strain energy density criterion applied to mixed-mode cracking dominated by in-plane shear

10.1 Preliminary remarks

Fracture mechanics is concerned with the assessment of material damage that involves both stable and unstable crack growth. The transition from one to another should be describable by a single failure criterion. The time dependent crack growth relation may be stated in the general form

$$\frac{da}{dt} = F(a, p, R). \tag{10.1}$$

The crack size a, load parameter p and fracture resistance parameter R that is characteristic of the material can, in general, be time dependent.

Crack growth resistance. A specific of equation (10.1) that describes the onset of crack growth by an amount Δa can be written as

$$F_R(a, p) \geqslant R(\Delta a). \tag{10.2}$$

The function $F_R(a, p)$ specifies the relation between a and p for crack growth; it can be associated with the strain energy density factor S [1–3]:

$$S = F_R. \tag{10.3}$$

Illustrated in Figure 10.1 is the crack growth resistance curve concept where the intersection of the R-curve and S-curve marks the transition from stable to unstable crack growth. This corresponds to a reaching the critical size a_c, i.e.,

$$\frac{\partial S}{\partial a} \geqslant \frac{\partial R}{\partial a} \quad \text{at } a = a_c. \tag{10.4}$$

G. C. Sih and E. E. Gdoutos (eds): *Mechanics and Physics of Energy Density*, 155–165.
© 1992 *Kluwer Academic Publishers. Printed in the Netherlands.*

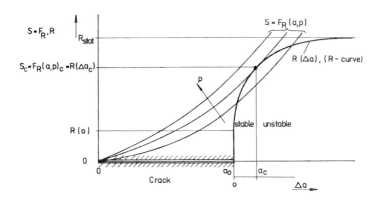

Figure 10.1. Crack growth resistance concept.

Equation (10.4) may be derived from equations (10.2) and (10.3) by assuming that $a = a(p)$:

$$\dot{a}\left(\frac{\partial S}{\partial a} - \frac{\partial R}{\partial a}\right) + \dot{p}\frac{\partial S}{\partial p} \geqslant 0. \tag{10.5}$$

Unstable crack growth corresponds to crack growth without further increase in load, i.e., $da > 0$ for $dp = 0$. The relation in equation (10.4) follows, therefore from equation (10.5) for $a = a_c$.

Onset of crack growth. The criterion of crack instability or the onset of unstable crack growth can thus be stated as

$$S_c(a, p) \geqslant R(\Delta a_c). \tag{10.6}$$

Referring to the strain energy density criterion in [1–4], the two fundamental hypotheses are

● Crack growth is assumed to occur in the direction θ_0 at which S attains a local minimum, i.e.,

$$\frac{\partial S}{\partial \theta} = 0, \quad \left(\frac{\partial^2 S}{\partial \theta^2}\right) > 0 \quad \text{for } \theta = \theta_0. \tag{10.7}$$

- Crack growth starts when $S_{\min}$ is assumed to reach a critical value S_c:

$$S_{\min}(a, p) = S_c, \tag{10.8}$$

such that both equations (10.4) and (10.6) are satisfied.

Stable crack growth. Stable crack growth has been discussed extensively in fracture mechanics concerning with cyclic loading. Many crack growth relations have been proposed, the majority of which can be expressed in the general form

$$\frac{da}{dt} = F_1(a)F_2(p). \tag{10.9}$$

A specific form of equation (10.9) is [2]

$$\frac{da}{dN} = C(\Delta S)^m, \tag{10.10}$$

where N is the number of loading cycles and C and m are material dependent parameters. The loading factors and effects of crack geometry are included in the strain energy density factor range ΔS. The values of C and m as evaluated from fatigue data are highly influenced by the ways with which ΔS are computed. That is ΔS computed from the theory of elasticity and plasticity would yield different values of C and m even for the same crack growth data, material, crack geometry and loading.

10.2 Sih's strain energy density criterion

Consider the isothermal linear elastic response of solids subjected to static loads. The well-known strain energy density criterion of Sih will be used to predict crack initiation under mixed-mode loading. It is expedient to introduce a modified stress intensity factor

$$K_m^* = K_I Q(\theta, M) \tag{10.11}$$

in which Q stands for

$$Q = \frac{1}{\sqrt{2(\kappa - 1)}} \left[a_{11} + a_{12} M + a_{22} M \right]^{\frac{1}{2}}. \tag{10.12}$$

For plane strain, $\kappa = 3 - 4v$ with v being Poisson's ratio. The quantity M is K_{II}/K_I where K_I and K_{II} are, respectively, the Mode I and II crack tip stress intensity factor. The coefficients a_{ij} in equation (10.12) are given by

$$a_{11} = (1 + \cos\ \theta)\ (\kappa - \cos\ \theta),$$

$$a_{12} = (2\ \cos\ \theta - \kappa + 1)\ \sin\ \theta, \tag{10.13}$$

$$a_{22} = 3\ \cos^2\theta + (1 - \kappa)\ \cos\ \theta + \kappa.$$

Application of equation (10.7) yields an equation for finding the direction of crack initiation θ_0:

$$\sin\ \theta_0\ (1 - \kappa + 2\ \cos\ \theta_0) + 2M\ [2\ \cos\ 2\theta_0 - (\kappa - 1)\ \cos\ \theta_0]$$

$$- M^2(1 - \kappa + 6\ \cos\ \theta_0)\ \sin\ \theta_0 = 0. \tag{10.14}$$

The envelope on which both K_I and K_{II} are critical is described by [5]:

$$\left(\frac{K_I}{K_{Ic}}\right)^2 a_{11}(\theta_0) + 2a_{12}(\theta_0) \left(\frac{K_I}{K_{Ic}}\right) \left(\frac{K_{II}}{K_{Ic}}\right) + a_{22}(\theta_0)$$

$$\left(\frac{K_{II}}{K_{IIc}}\right)^2 \frac{1}{2(\kappa - 1)} = 1. \tag{10.15}$$

If a tensile load is directed at an angle α with the x-axis in Figure 10.2, then the crack would initiate in some direction θ_0 that is not necessarily coincident

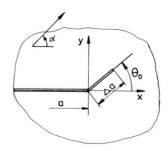

Figure 10.2. Angled crack growth.

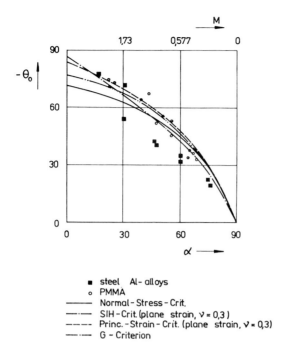

Figure 10.3.   Crack propagation angle versus load angle (under tensile load).

with the x-axis. The results of equation (10.14) are displayed in Figure 10.3 with $-\theta_0$ plotted as a function of α for different M. Experimental results for steel, aluminum and PMMA are also shown. A plot of K_{II}/K_{IIc} versus K_I/K_{Ic} according to equation (10.15) is shown in Figure 10.4 for $v = 0.3$. The other three curves are obtained from the maximum normal stress, principal strain and energy release rate criterion. Experimental data for glass, PMMA, and aluminum [6] show that no conclusion could be drawn as they tend to spread among all of the predictions. To this end, a modification would be made in the work to follow.

10.3 Mixed-mode cracking dominated by in-plane shear

Let $K_I^c = K_{Ic}$ and $K_{II}^c = K_{IIc}$ be defined, respectively, as the cases where the critical stress intensity factors are obtained from the *pure* Mode I and II crack extension as in [7]. Introduce the relation

$$K_{Ic} = \frac{K_{II}^c}{\varphi_s} \quad \text{or} \quad K_{Ic} = \frac{K_{IIc}}{\varphi_s}. \tag{10.16}$$

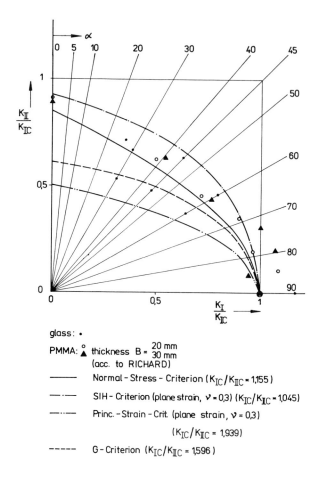

glass: •
PMMA: ○ thickness B = 20 mm / 30 mm (acc. to RICHARD)
▲
——— Normal - Stress - Criterion (K_{IC}/K_{IIC} = 1,155)
— ·— SIH - Criterion (plane strain, ν = 0,3) (K_{IC}/K_{IC} = 1,045)
— ··— Princ. - Strain - Crit. (plane strain, ν = 0,3)
                   (K_{IC}/K_{IC} = 1,939)
- - - - G - Criterion (K_{IC}/K_{IIC} = 1,596)

Figure 10.4   Fracture envelope in the K_1/K_{1c} versus K_{11}/K_{1c} plane.

Alternatively, equation (10.16) may be written as

$$K_{1c} = K_{11}^c \left(\frac{\beta_s}{\varphi_s}\right) \tag{10.17}$$

provided that

$$\beta_s = \varphi_s\left(\frac{K_{1c}}{K_{11c}}\right). \tag{10.18}$$

Mode II crack extension. When $K_1 = 0$ and $K_{11} \neq 0$, the modified stress intensity factor becomes

$$K_m^* = \frac{\sqrt{a_{22}}}{\sqrt{2(\kappa - 1)}} K_{\mathrm{II}}.$$

(10.19)

The angle of crack initiation can be obtained by applying equation (10.7)

$$\frac{\partial a_{22}}{\partial \theta} = 0, \qquad \frac{\partial^2 a_{22}}{\partial \theta^2} > 0 \qquad \text{for } \theta = \theta_0.$$

(10.20)

This gives

$$\theta_0 = -\cos^{-1}\left(\frac{\kappa - 1}{6}\right)$$

(10.21)

and, hence,

$$a_{22}(\theta_0) = \kappa - \frac{(\kappa - 1)^2}{12}.$$

(10.22)

At the onset of crack instability, equation (10.19) becomes

$$(K_m^*)_c = \frac{\sqrt{a_{22}(\theta_0)}}{\sqrt{2(\kappa - 1)}} K_{\mathrm{II}}^c = K_{\mathrm{I}c}.$$

(10.23)

It follows from equation (10.16) that

$$\varphi_s = \left[\frac{24(\kappa - 1)}{12\kappa - (\kappa - 1)}\right]^{1/2}.$$

(10.24)

Modified stress intensity factor. The modified stress intensity factor K_m^* for Mode II crack extension may be obtained if K_{II} and M are replaced, respectively, by $K_{\mathrm{II}}\beta_s$ and $M^* = M\beta_s$. In relation to the Sih criterion, K_m^* takes the form

$$K_m^* = K_{\mathrm{I}} Q^*(\theta, M^*),$$

(10.25)

in which

$$Q^* = \frac{1}{\sqrt{2(\kappa - 1)}} [a_{11} + 2a_{12}M^* + a_{22}(M^*)^2]^{\frac{1}{2}}.$$

(10.26)

The angle θ_0 may be calculated from equation (10.14) if M is replaced by M^*.

The Sih criterion can be used such that the ratio K_{Ic}/K_{IIc} may be made analogous to the normal stress criterion, the principal strain criterion and the energy volume rate criterion and any others.

Normalized Mode I and II stress intensity factor. The representation of fracture data on the plane K_I/K_{Ic} versus K_{II}/K_{Ic} requires the specification of the ratio K_{Ic}/K_{IIc}. It is more expedient to present the data in terms of the normalized quantities K_I/K_{Ic} and K_{II}/K_{IIc}. For each value of M^*, the values

$$\frac{K_I^c}{K_{Ic}} = \frac{1}{Q^*}, \qquad \frac{K_{II}^c}{K_{IIc}} = \frac{M}{SQ^*} \tag{10.27}$$

may be computed to establish the envelope in Figure 10.5. A parabolic relation

$$\frac{K_I^c}{K_{Ic}} + \left(\frac{|K_{II}^c|}{K_{IIc}}\right)^\eta = 1 \tag{10.28}$$

may be used. For $K_I^c = 0$ and $v = 0.3$, the exponent $\eta \simeq 2.5$. The influence of M^* on the fracture envelope can be noted from

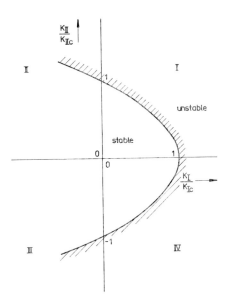

Figure 10.5. Fracture envelope based on normalized Mode I and II stress intensity factor.

$$M^* = M \, \varphi_s \left(\frac{K_{Ic}}{K_{IIc}} \right) \tag{10.29}$$

which includes the influence of geometry and loading. Here, φ_s describes the effect of Mode II crack extension.

Continuation of the fracture envelope into the II and III quadrant shown in Figure 10.5 involve negative K_1 values. This corresponds to crack closure, an effect that has not been investigated.

10.4 Discussions

Fracture data have been interpreted in terms of the ratio K_{Ic} and K_{IIc}. Fracture angle θ_0^* obtained using M^* versus K_{II}/K_I or M relation changes for each K_{Ic}/K_{IIc} ratio as shown in Figure 10.6. The same applies to the plot of K_{II}/K_{Ic} against K_I/K_{Ic}; this is displayed in Figure 10.7 together with the experimental data on Araldit B [6].

Comparison of Mode II crack extension data with prediction is often hampered by the presence of crack surface friction effects that are not corrected in the analysis. If μ^* stands for the coefficient of friction, a correction on K_{IIc} can be made such that

$$(K_{IIc})_f = (1 + \mu^*) K_{IIc}. \tag{10.30}$$

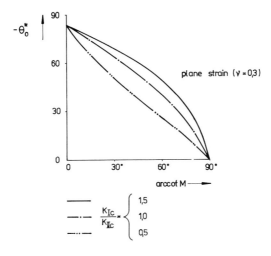

Figure 10.6.   Crack initiation angle versus M factor for the Sih criterion.

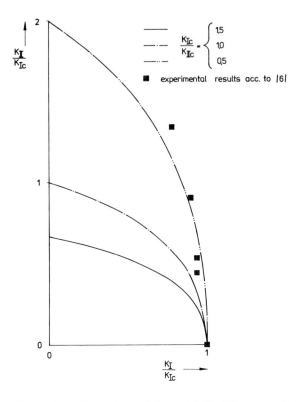

Figure 10.7. Comparison of experimental data with K_{II}/K_{Ic} versus K_I/K_{Ic} plots.

It follows that

$$\frac{K_{Ic}}{K_{IIc}} > \left(\frac{K_{Ic}}{K_{IIc}}\right)_{exp}. \tag{10.31}$$

This would lead to an enlargement of the fracture envelope in Figure 10.7. A possible explanation for the difference between the experiment data for Araldit B with $K_{Ic}/K_{IIc} = 0.3$ in [6] is thus offered when Mode II crack extension dominates.

References

[1] Sih, G.C., "A special theory of crack propagation", in: *Mechanics of Fracture* 1 (Ed.: G.C. Sih), Noordhoff Int. Publ., Leyden (1973).
[2] Sih, G.C. and Moyer, Jr., E.T., "Path dependent nature of fatigue crack growth", *Eng. Fract. Mech.*, **17**, pp. 269–280 (1983).

[3] Sih, G.C., "Mechanics and physics of energy density theory", *Theoret. and Appl. Fract. Mech.,* **4** No. 3, pp. 157–173 (1985).

[4] Fischer, K.-F., Bruchkriterien bei inhomogener Beanspruchung", in: *Beurteilungs- und Bruchkriterien,* VEB Fachbuchverlag Leipzig (1989).

[5] Fischer, K.-F., "Critique on fracture criteria in mixed-mode-loading", *Theoret. and Appl. Fract. Mech., Vol.* **3**, pp. 85–95 (1985).

[6] Kordisch, H., Richard, H.A., and Sommer, E., private communication (1983).

[7] Irwin, G.R., "Analysis of stresses and strains near the end of a crack traversing plate", *Appl. Mech.,* **24** pp. 361–364 (1957).

Group-averaging methods for generating constitutive equations

11.1 Introduction

We are concerned with the problem of determining the form of constitutive equations which define the non-linear response of materials possessing symmetry properties. This requires that we find the canonical form of scalar-valued functions $W(\mathbf{E}, \mathbf{F}, ...)$ and tensor-valued functions $\mathbf{T}(\mathbf{E}, \mathbf{F}, ...)$ which are invariant under the group of transformations $\{A\}$ defining the material symmetry. Solution of these problems frequently requires a substantial computational effort. We have initiated in [1] an approach designed to yield a family of computer programs which will enable us to deal with the invariant-theoretic problems associated with the formulation of constitutive equations. We consider here an approach based upon averaging over the symmetry group $\{A\}$ which is well adapted to computer based computation. We limit consideration in this paper to the case where $W = W(\mathbf{E})$ and $\mathbf{T} = \mathbf{T}(\mathbf{E})$ are functions of a single tensor $\mathbf{E}$. We further limit consideration to those cases where the symmetry group $\{A\}$ is a finite group. In 11.2, we discuss the manner in which sets of linearly independent scalar-valued invariants of degrees 1, 2, ... in $\mathbf{E}$ may be generated. In 11.3, we outline the procedure for generating sets of linearly independent tensor-valued invariant functions of degrees 1, 2, ... in $\mathbf{E}$. In 11.4, we give examples to illustrate the application of the results given in 11.2 and 11.3. A discussion of the group theoretic notions employed below may be found in Murnaghan [2] or Lomont [3].

11.2 Generation of scalar-valued invariants

We consider the problem of determining the form of a scalar-valued function $W(E_{p...q})$, of given degree in the components of a tensor $\mathbf{E} = E_{p...q}$, which is

G. C. Sih and E. E. Gdoutos (eds): Mechanics and Physics of Energy Density, 167–178.

invariant under a group $\{A\} = \{\mathbf{A}_1, ..., \mathbf{A}_N\}$. Thus, $W(E_{p...q})$ must satisfy

$$W(E_{p...q}) = W(A^K_{pr...}\, A^K_{qs}\, E_{r...s}) \tag{11.1}$$

for all $\mathbf{A}_K = \| A^K_{pr} \|$ belonging to $\{A\}$. Let $E_1, ..., E_r$ denote the independent components of $\mathbf{E}$. We then consider the equivalent problem of determining the form of $W(E_i) = W(\mathbf{E})$ where E denotes the column matrix $\| E_1, ..., E_r \|^T$. The restrictions corresponding to (11.1) are that $W(\mathbf{E})$ must satisfy

$$W(\mathbf{E}) = W(\mathbf{R}_K \mathbf{E}) \tag{11.2}$$

for $K = 1, ..., N$ where $R_K, (K = 1, ..., N)$ is the $r \times r$ matrix representation of the group $\{A\}$ which defines the transformation properties of $\mathbf{E}$ under $\{A\}$. We first consider the special case where W is a linear function of the components of E. We observe that the r entries of the column vector

$$\mathbf{R}_1 \mathbf{E} + ... + \mathbf{R}_N \mathbf{E} = \sum_{K=1}^{N} R^K_{ij} E_j \qquad (i, j = 1, ..., r) \tag{11.3}$$

are either invariants or zeros. Thus replacing E by $\mathbf{R}_K \mathbf{E}$ in equation (11.3) leaves the expression unaltered since

$\mathbf{R}_1 \mathbf{R}_K + ... + \mathbf{R}_N \mathbf{R}_K = \mathbf{R}_1 + \cdots + \mathbf{R}_N$ for any $\mathbf{R}_K$. We employ the notation

$$\mathbf{R}^*_1 = \| \sum_{K=1}^{N} R^K_{ij} \|. \tag{11.4}$$

Each row of this matrix, when multiplied by the column vector E, yields either an invariant or zero. The number of linearly independent invariants of degree 1 in **E** is given by

$$n_1 = \frac{1}{N} \operatorname{tr} \mathbf{R}^*_1 = \frac{1}{N} \sum_{K=1}^{N} \operatorname{tr} \mathbf{R}_K. \tag{11.5}$$

We may carry out a row reduction procedure on the matrix $\mathbf{R}^*_1$ which has rank n_1 to obtain a matrix $\mathbf{S}_1$ with n_1 rows which gives a set of n_1 linearly independent invariants when multiplied by the column matrix E.

We may generate the invariants of degree 2 in a similar fashion. Let $\mathbf{E}^2$ denote the column matrix

$$\mathbf{E}^2 = \| E_1^2, E_1 E_2, ..., E_1 E_r, E_2^2, E_2 E_3, ..., E_2 E_r, ..., E_{r-1}^2, E_{r-1} E_r, E_r^2 \|^T \quad (11.6)$$

whose entries are the $\frac{1}{2}r(r+1)$ distinct monomials of degree two in the $E_1, ...,$ E_r. Let $\mathbf{R}_K^{(2)}$, $(K = 1, ..., N)$ denote the matrix representation of $\{A\}$ which defines the transformation properties of $\mathbf{E}^2$ under $\{A\}$. $\mathbf{R}_K^{(2)}$ is a $\frac{1}{2}r(r+1)$ $\times \frac{1}{2}r(r+1)$ matrix which is referred to as the symmetrized Kronecker square of $\mathbf{R}_K$. Then

$$\mathbf{R}_2^* = \sum_{K=1}^{N} \mathbf{R}_K^{(2)},$$

$$n_2 = \frac{1}{N} \operatorname{tr} \mathbf{R}_2^* = \frac{1}{N} \sum_{K=1}^{N} \operatorname{tr} \mathbf{R}_K^{(2)}, \qquad (11.7)$$

where $\mathbf{R}_2^*$ is a matrix whose rows yield invariants (or zero) when multiplied by the column matrix $\mathbf{E}^2$ and where n_2 gives the number of linearly independent invariants of degree 2 in $\mathbf{E}$. The rank of the matrix $\mathbf{R}_2^*$ is n_2 and we apply a row reduction procedure to $\mathbf{R}_2^*$ to obtain a matrix $\mathbf{S}_2$ which has n_2 linearly independent rows. Multiplication of each row by the column matrix $\mathbf{E}^2$ yields a set of n_2 linearly independent invariants.

We may continue this process and generate invariants of any given degree n in $\mathbf{E}$. We denote by $\mathbf{E}^n$ the column matrix

$$\mathbf{E}^n = \| E_1^n, E_1^{n-1} E_2, E_1^{n-1} E_3, ..., E_r^n \|^T \qquad (11.8)$$

whose entries are the $\binom{r+n-1}{n}$ distinct monomials of degree n in $E_1, ..., E^r$. The set of N matrices $\mathbf{R}_K^{(n)}$ $(K = 1, ..., N)$, forms a matrix representation of $\{A\}$ which defines the transformation properties of $\mathbf{E}^n$. The matrix $\mathbf{R}_K^{(n)}$ is referred to as the symmetrized Kronecker n^{th} power of $\mathbf{R}_K$. Then

$$\mathbf{R}_n^* = \sum_{K=1}^{N} \mathbf{R}_K^{(n)} \qquad (11.9)$$

is a matrix whose rows yield invariants (or zero) when multiplied by the column matrix $\mathbf{E}^n$. The number of linearly independent invariants of degree n in $\mathbf{E}$ is given by

$$n_n = \frac{1}{N} \operatorname{tr} \mathbf{R}_n^* = \frac{1}{N} \sum_{K=1}^{N} \operatorname{tr} \mathbf{R}_K^{(n)}. \qquad (11.10)$$

The rank of $\mathbf{R}_n^*$ is n_n. Applying a row reduction procedure to $\mathbf{R}_n^*$ yields a matrix $\mathbf{S}_n$ which has n_n linearly independent rows. Multiplication of each row by the column matrix $\mathbf{E}^n$ yields a set of n_n linearly independent invariants.

Thus, in order to generate the invariants of any given degree n in $\mathbf{E}$, we need only be able to produce the set of matrices $\mathbf{R}_K^{(n)}$. The entries in the matrix $\mathbf{R}_K^{(n)}$ are polynomial functions of the R_{ij}^K. We note that if $\mathbf{R}_K$ is a 4×4 matrix, then $\mathbf{R}_K^{(6)}$ is an 84×84 matrix. Consequently, computation of the $\mathbf{R}_K^{(n)}$ for large n would be tedious and error prone. To avoid these difficulties, we are writing a computer program which will produce the symmetrized Kronecker n^{th} power $\mathbf{R}_K^{(n)}$ of a matrix $\mathbf{R}_K$. We combine this program with a row reduction subroutine which enables us to automatically generate the matrices $\mathbf{S}_1, \mathbf{S}_2, ...,$ $\mathbf{S}_n$ and hence to automatically generate sets of linearly independent invariants of any given order in $\mathbf{E}$.

11.3 Generation of tensor-valued invariant functions

We consider the problem of determining the form of a tensor-valued function

$$T_{i...j} = F_{i...j}(E_{p...q}) \tag{11.11}$$

of given degree in the components of $\mathbf{E}$ which is invariant under a group $\{A\}$. The invariance requirement is that

$$A_{ik}^K ... A_{jl}^K F_{k...l}(E_{p...q}) = F_{i...j}(A_{pr...}^K A_{qs}^K E_{r...s}) \tag{11.12}$$

must hold for all $\mathbf{A}_K$ belonging to $\{A\}$. Let $T_1, ..., T_s$ and $E_1, ..., E_r$ denote the independent components of $\mathbf{T}$ and $\mathbf{E}$ respectively. We may then write (11.11) as

$$T_i = F_i(E_j) \quad \text{or} \quad \mathbf{T} = \mathbf{F}(\mathbf{E}), \tag{11.13}$$

where $\mathbf{T}$ and $\mathbf{E}$ denote the column vectors

$$\mathbf{T} = \| T_1, ..., T_s \|^T, \qquad \mathbf{E} = \| E_1, ..., E_r \|^T. \tag{11.14}$$

If the expressions in equation (11.13) are to be invariant under $\{A\}$, then

$$S_{ij}^K F_j(E_k) = F_i(R_{kl}^K E_l) \quad \text{or} \quad \mathbf{S}_K \mathbf{F}(\mathbf{E}) = \mathbf{F}(\mathbf{R}_K \mathbf{E}) \tag{11.15}$$

must hold for $K = 1, ..., N$ where $\mathbf{S}_K$ and $\mathbf{R}_K$, $(K = 1, ..., N)$ are the matrix representations of $\{A\}$ which define the transformation properties of $\mathbf{T}$ and $\mathbf{E}$ respectively under $\{A\}$. Let

$$T_i^* = P_{ij} T_j, (i, j = 1, ..., s) \quad \text{or} \quad \mathbf{T}^* = \mathbf{PT}, \tag{11.16}$$

where **T*** is a column vector whose entries are linear combinations of $T_1, ..., T_s$. The matrix representation defining the manner in which **T*** transforms under $\{A\}$ is given by

$$\mathbf{S}_K^* = \mathbf{P}\mathbf{S}_K\mathbf{P}^{-1}, \quad (K = 1, ..., N). \tag{11.17}$$

We may choose **P** so that the matrix representation $\mathbf{S}_K, (K = 1, ..., N)$, is decomposed into the direct sum of irreducible representations of $\{A\}$, i.e.

$$\mathbf{S}_K^* = \mathbf{P}\mathbf{S}_K\mathbf{P}^{-1} = \alpha_1\,\mathbf{D}_K^{(1)} \dot{+} \alpha_2\,\mathbf{D}_K^{(2)} \dot{+} ... \dot{+} \alpha_a\mathbf{D}_K^{(a)}, \quad (K = 1, ..., N) \tag{11.18}$$

where for example

$$2\mathbf{D}_K^{(1)} + \mathbf{D}_K^{(2)} = \begin{Vmatrix} \mathbf{D}_K^{(1)} & 0 & 0 \\ 0 & \mathbf{D}_K^{(1)} & 0 \\ 0 & 0 & \mathbf{D}_K^{(2)} \end{Vmatrix}. \tag{11.19}$$

Thus we may split the s components of **T** into $\alpha_1 + \alpha_2 + ... + \alpha_a$ sets of components $(T^*, T_2^*), (T_3^*, T_4^*, T_5^*),...$ whose transformation properties under $\{A\}$ are described by the irreducible representations $\mathbf{D}_K^{(1)}, ..., \mathbf{D}_K^{(a)}$. This reduces the problem of determining the form of $\mathbf{T} = \mathbf{F}(\mathbf{E})$ which is consistent with the restriction (11.15) to the set of simpler problems of determining the forms of

$$\mathbf{T}^{(k)} = \mathbf{F}_k(\mathbf{E}), \; (k = 1, ..., a) \tag{11.20}$$

which is subject to the restrictions that

$$\mathbf{D}_K^{(k)}\mathbf{F}_k(\mathbf{E}) = \mathbf{F}_k(\mathbf{R}_K\mathbf{E}), \quad (K = 1, ..., N). \tag{11.21}$$

In the remainder of the paper, we shall be concerned then only with problems of determining the form of functions

$$\mathbf{T} = \mathbf{F}(\mathbf{E}) \tag{11.22}$$

which are subject to the restrictions

$$\mathbf{D}_K\mathbf{F}(\mathbf{E}) = \mathbf{F}(\mathbf{R}_K\mathbf{E}), \quad (K = 1, ..., N) \tag{11.23}$$

where the matrices $\mathbf{D}_K, (K = 1, ..., N)$, form an *irreducible* matrix representation of the group $\{A\}$. We may (and shall) assume with no loss of generality that the matrices $\mathbf{D}_K$ are unitary, i.e.

$$\mathbf{D}_K^{-1} = \overline{\mathbf{D}}_K^T, \qquad (\mathbf{D}_K^{-1})_{ij} = \overline{\mathbf{D}}_{ji}^K \tag{11.24}$$

where the superposed bar indicates the complex conjugate.

Consider the special case where the functions $F_m(\mathbf{E})$ appearing in equations (11.22) and (11.23) are linear. We observe that

$$F_m(\mathbf{E}) = \sum_{K=1}^{N} (\mathbf{D}_K^{-1})_{mi} (\mathbf{R}_K)_{jp} E_p, \quad (m = 1, ..., s), \tag{11.25}$$

satisfies equation (11.23) for each value of i and j. Thus, replacing E_p by $(\mathbf{R}_L)_{pq} E_q$ on the right of (11.25) yields

$$\sum_{K=1}^{N} (\mathbf{D}_K^{-1})_{mi} (\mathbf{R}_K)_{jp} (\mathbf{R}_L)_{pq} E_q$$

$$= \sum_{M=1}^{N} (\mathbf{D}_L \mathbf{D}_M^{-1})_{mi} (\mathbf{R}_M)_{jq} E_q \tag{11.26}$$

$$= (\mathbf{D}_L)_{mn} \sum_{M=1}^{N} (\mathbf{D}_M^{-1})_{ni} (\mathbf{R}_M)_{jp} E_p = (\mathbf{D}_L)_{mn} F_n(\mathbf{E})$$

where we have set $\mathbf{R}_K \mathbf{R}_L = \mathbf{R}_M$, $\mathbf{D}_K \mathbf{D}_L = \mathbf{D}_M$, $\mathbf{D}_K = \mathbf{D}_M \mathbf{D}_L^{-1}$ and $\mathbf{D}_K^{-1} = \mathbf{D}_L \mathbf{D}_M^{-1}$. With equations (11.24) and (11.25), we may write

$$F_m(\mathbf{E}) = \sum_{K=1}^{N} \overline{D}_{im}^K R_{jp}^K E_p, \quad (i, m = 1, ..., s; j, p = 1, ..., r). \tag{11.27}$$

Thus, setting $i = 1, ..., s$ and $j = 1, ..., r$ in turn in equations (11.27), we may obtain rs functions $F_m(\mathbf{E})$, $(m = 1, ..., s)$, which are invariant under $\{A\}$. There are only

$$n_1^* = \frac{1}{N} \sum_{K=1}^{N} \operatorname{tr} \overline{\mathbf{D}}_K \operatorname{tr} \mathbf{R}_K \tag{11.28}$$

of these functions which are linearly independent. We note that for fixed values of $(i, j) = (1, 1)$ say in equation (11.27) the sr quantities $\sum_{K=1}^{N} \overline{D}_{lm}^K R_{lp}^K, (m = 1, ..., s; p = 1, ..., r)$ form an $s \times r$ matrix (which may consist of all zeros). Appropriate choice of i, j in the expressions $\sum_{K=1}^{N} \overline{D}_{im}^K R_{jp}^K$ will yield n_1^* sets of $s \times r$ matrices. We may employ a row reduction procedure to determine that the $n_1^* s$ row vectors occurring are linearly independent. Multiplication of each $s \times r$ matrix obtained for each of the n_1^* particular choices of (i, j) will yield upon multiplication by the column matrix $\mathbf{E}$ a function $\mathbf{F}(\mathbf{E})$ which satisfies $\mathbf{D}_K \mathbf{F}(\mathbf{E}) = \mathbf{F}(\mathbf{R}_K \mathbf{E})$.

We may generate functions $\mathbf{F}(\mathbf{E})$ which satisfy equation (11.23) and which are of degree 2 in the components of $\mathbf{E}$ in a similar fashion. Thus let $\mathbf{E}^2$ denote the column vector defined by equation (11.6) and let $\mathbf{R}_K^{(2)}$ denote the symmetrized Kronecker square of $\mathbf{R}_K$. There are n_2^* sets of linearly independent functions $\mathbf{F}(\mathbf{E})$ of degree 2 in $\mathbf{E}$ which satisfy equation (11.23) where

$$n_2^* = \frac{1}{N} \sum_{K=1}^{N} \operatorname{tr} \bar{\mathbf{D}}_K \operatorname{tr} \mathbf{R}_K^{(2)}. \tag{11.29}$$

We may generate these n_2^* functions from the expression

$$\sum_{K=1}^{N} \bar{D}_{im}^K R_{jp}^{(2)K} E_p^2 \left(i, m = 1, ..., s; j, p = 1, ..., \binom{r+1}{2} \right), \tag{11.30}$$

where $\mathbf{E}^2$ is defined by equation (11.6). Thus, for fixed values of (i, j), e.g. $(i, j) = (1, 1)$, the $s \times \binom{r+1}{2}$ matrix $\Sigma_{K=1}^{N} \bar{D}_{im}^K R_{ip}^{(2)K}$ will yield a function $\mathbf{F}(\mathbf{E})$ of degree 2 in $\mathbf{E}$ which satisfies equation (11.23). Appropriate choices of values for (i, j) will then yield n_2^* linearly independent functions $\mathbf{F}(\mathbf{E})$ which satisfy equation (11.23). This process may be employed to generate functions $\mathbf{F}(\mathbf{E})$ of any degree n in $\mathbf{E}$. We need only replace $\mathbf{E}^2$ by $\mathbf{E}^n$ and $\mathbf{R}_K^{(2)}$ by $\mathbf{R}_K^{(n)}$ in equation (11.30). The number n_n^* of sets of linearly independent functions of degree n in $\mathbf{E}$ is given by

$$n_n^* = \frac{1}{N} \sum_{K=1}^{N} \operatorname{tr} \bar{\mathbf{D}}_K \operatorname{tr} \mathbf{R}_K^{(n)}. \tag{11.31}$$

We may generate these n_n^* functions from the expression

$$\sum_{K=1}^{N} \bar{D}_{im}^K R_{jp}^{(n)K} E_p^n \left(i, m = 1, ..., s; j, p = 1, ..., \binom{r+n-1}{n} \right), \tag{11.32}$$

where $\mathbf{E}^n$ is defined by equation (11.8).

11.4 Applications

In this section, we give some examples of the application of the results of the previous sections to the determination of functions which are invariant under a group $\{A\} = \{\mathbf{A}_1, \mathbf{A}_2, ..., \mathbf{A}_6\}$ where

$$\mathbf{A}_1 = \left\| \begin{matrix} 1, & 0 \\ 0, & 1 \end{matrix} \right\|, \quad \mathbf{A}_2 = \left\| \begin{matrix} -1/2, & \sqrt{3}/2 \\ -\sqrt{3}/2, & -1/2 \end{matrix} \right\|, \quad \mathbf{A}_3 = \left\| \begin{matrix} -1/2, & -\sqrt{3}/2 \\ \sqrt{3}/2, & -1/2 \end{matrix} \right\|,$$

$$\mathbf{A}_4 = \left\| \begin{matrix} -1, & 0 \\ 0, & 1 \end{matrix} \right\|, \quad \mathbf{A}_5 = \left\| \begin{matrix} 1/2, & -\sqrt{3}/2 \\ -\sqrt{3}/2, & -1/2 \end{matrix} \right\|, \quad \mathbf{A}_6 = \left\| \begin{matrix} 1/2, & \sqrt{3}/2 \\ \sqrt{3}/2, & -1/2 \end{matrix} \right\|. \tag{11.33}$$

We may employ the method of 11.2 to generate scalar-valued functions $W(\mathbf{x})$ of a two-dimensional vector $\mathbf{x} = \| x_1, x_2 \|^T$ which are invariant under the group $\{A\}$. The matrix representation $\mathbf{R}_K$, $(K = 1, ..., 6)$, which defines the transformation properties of $\mathbf{x}$ under $\{A\}$ is given by

$$\mathbf{R}_K = \mathbf{A}_K, \ (K = 1, ..., 6). \tag{11.34}$$

With equations (11.5), (11.33) and (11.34), we see that the number of invariants of degree one in $\mathbf{x}$ is given by $n_1 = \dfrac{1}{6}(\operatorname{tr} \mathbf{A}_1 + ... + \operatorname{tr} \mathbf{A}_6) = 0$. In order to determine the invariants of degree two in $\mathbf{x}$, we list the symmetrized Kronecker squares $\mathbf{A}_K^{(2)}$ of the $\mathbf{A}_K$. These are given by

$$\mathbf{A}_1^{(2)} = \begin{Vmatrix} 1, & 0, & 0 \\ 0, & 1, & 0 \\ 0, & 0, & 1 \end{Vmatrix}, \quad \mathbf{A}_2^{(2)} = \begin{Vmatrix} 1/4, & -\sqrt{3}/2, & 3/4 \\ \sqrt{3}/4, & -1/2, & -\sqrt{3}/4 \\ 3/4, & \sqrt{3}/2, & 1/4 \end{Vmatrix},$$

$$\mathbf{A}_3^{(2)} = \begin{Vmatrix} 1/4, & \sqrt{3}/2, & 3/4 \\ -\sqrt{3}/4, & -1/2, & \sqrt{3}/4 \\ 3/4, & -\sqrt{3}/2, & 1/4 \end{Vmatrix}, \quad \mathbf{A}_4^{(2)} = \begin{Vmatrix} 1, & 0, & 0 \\ 0, & -1, & 0 \\ 0, & 0, & 1 \end{Vmatrix}, \tag{11.35}$$

$$\mathbf{A}_5^{(2)} = \begin{Vmatrix} 1/4, & -\sqrt{3}/2, & 3/4 \\ -\sqrt{3}/4, & 1/2, & \sqrt{3}/4 \\ 3/4, & \sqrt{3}/2, & 1/4 \end{Vmatrix}, \quad \mathbf{A}_6^{(2)} = \begin{Vmatrix} 1/4, & \sqrt{3}/2, & 3/4 \\ \sqrt{3}/4, & 1/2, & -\sqrt{3}/4 \\ 3/4, & -\sqrt{3}/2, & 1/4 \end{Vmatrix}.$$

We note that the matrices $\mathbf{A}_K^{(2)}$ define the manner in which the quantity $\mathbf{x}^2$ transforms under $\{A\}$ where

$$\mathbf{x}^2 = \begin{Vmatrix} x_1^2 \\ x_1 x_2 \\ x_2^2 \end{Vmatrix}. \tag{11.36}$$

With equations (11.7) and (11.35), we see that the number of linearly independent invariants of degree 2 in $\mathbf{x}$ is given by

$$n_2 = \frac{1}{6}(\operatorname{tr} \mathbf{A}_1^{(2)} + ... + \operatorname{tr} \mathbf{A}_6^{(2)}) = 1. \tag{11.37}$$

With equations (11.7) and (11.35), we see that

$$\sum_{K=1}^{6} \mathbf{A}_K^{(2)}\mathbf{x}^2 = \begin{Vmatrix} 3, & 0, & 3 \\ 0, & 0, & 0 \\ 3, & 0, & 3 \end{Vmatrix} \begin{Vmatrix} x_1^2 \\ x_1 x_2 \\ x_2^2 \end{Vmatrix} = 3 \begin{Vmatrix} x_1^2 + x_2^2 \\ 0 \\ x_1^2 + x_2^2 \end{Vmatrix} \tag{11.38}$$

which yields the result that $x_1^2 + x_2^2$ is an invariant. We next list the symmetrized Kronecker cubes $\mathbf{A}_K^{(3)}$ of the $\mathbf{A}_K$.

$$\mathbf{A}_1^{(3)} = \begin{Vmatrix} 1, & 0, & 0, & 0 \\ 0, & 1, & 0, & 0 \\ 0, & 0, & 1, & 0 \\ 0, & 0, & 0, & 1 \end{Vmatrix}, \quad \mathbf{A}_2^{(3)} = \begin{Vmatrix} -1/8, & 3\sqrt{3}/8, & -9/8, & 3\sqrt{3}/8 \\ -\sqrt{3}/8, & 5/8, & -\sqrt{3}/8, & -3/8 \\ -3/8, & \sqrt{3}/8, & 5/8, & \sqrt{3}/8 \\ -3\sqrt{3}/8, & -9/8, & -3\sqrt{3}/8, & -1/8 \end{Vmatrix},$$

$$\mathbf{A}_3^{(3)} = \begin{Vmatrix} -1/8, & -3\sqrt{3}/8, & -9/8, & -3\sqrt{3}/8 \\ \sqrt{3}/8, & 5/8, & \sqrt{3}/8, & -3/8 \\ -3/8, & -\sqrt{3}/8, & 5/8, & -\sqrt{3}/8 \\ 3\sqrt{3}/8, & -9/8, & 3\sqrt{3}/8, & -1/8 \end{Vmatrix}, \quad \mathbf{A}_4^{(3)} = \begin{Vmatrix} -1, & 0, & 0, & 0 \\ 0, & 1, & 0, & 0 \\ 0, & 0, & -1, & 0 \\ 0, & 0, & 0, & 1 \end{Vmatrix}, \tag{11.39}$$

$$\mathbf{A}_5^{(3)} = \begin{Vmatrix} 1/8, & -3\sqrt{3}/8, & 9/8, & -3\sqrt{3}/8 \\ -\sqrt{3}/8, & 5/8, & -\sqrt{3}/8, & -3/8 \\ 3/8, & -\sqrt{3}/8, & -5/8, & -\sqrt{3}/8 \\ -3\sqrt{3}/8, & -9/8, & -3\sqrt{3}/8, & -1/8 \end{Vmatrix},$$

$$\mathbf{A}_6^{(3)} = \begin{Vmatrix} 1/8, & 3\sqrt{3}/8, & 9/8, & 3\sqrt{3}/8 \\ \sqrt{3}/8, & 5/8, & \sqrt{3}/8, & -3/8 \\ 3/8, & \sqrt{3}/8, & -5/8, & \sqrt{3}/8 \\ 3\sqrt{3}/8, & -9/8, & 3\sqrt{3}/8, & -1/8 \end{Vmatrix}.$$

With equations (11.10) and (11.39), we have that the number of linearly independent invariants of degree 3 in $\mathbf{x}$ is given by

$$n_3 = \frac{1}{6}(\operatorname{tr}\mathbf{A}_1^{(3)} + \dots + \operatorname{tr}\mathbf{A}_6^{(3)}) = 1. \tag{11.40}$$

With (11.39), we see that

$$\sum_{K=1}^{6} \mathbf{A}_K^{(3)} \mathbf{x}^3 = \begin{Vmatrix} 0, & 0, & 0, & 0 \\ 0, & 9/2, & 0, & -3/2 \\ 0, & 0, & 0, & 0 \\ 0, & -9/2, & 0, & 3/2 \end{Vmatrix} \begin{Vmatrix} x_1^3 \\ x_1^2 x_2 \\ x_1 x_2^2 \\ x_2^3 \end{Vmatrix} \tag{11.41}$$

which yields the result that $(3x_1^2 - x_2^2)x_2$ is an invariant.

We next consider the problem of determining the form of a vector-valued function $\mathbf{z}(\mathbf{x})$ which is invariant under $\{A\} = \{\mathbf{A}_1, ..., \mathbf{A}_6\}$. The matrix representation $\mathbf{R}_K$ which defines the transformation properties of $\mathbf{z} = \|z_1, z_2\|^T$ under $\{A\}$ is given by

$$\mathbf{R}_K = \mathbf{A}_K, \quad (K = 1, ..., 6), \tag{11.42}$$

which is of course the same as for the vector $\mathbf{x}$. We note that there are three inequivalent irreducible representations of the group $\{A\} = \{\mathbf{A}_1, ..., \mathbf{A}_6\}$ which are given by

$$\begin{aligned} &\Gamma_1(\mathbf{A}_1), ..., \Gamma_1(\mathbf{A}_6) = 1, 1, 1, 1, 1, 1; \\ &\Gamma_2(\mathbf{A}_1), ..., \Gamma_2(\mathbf{A}_6) = 1, 1, 1, -1, -1, -1; \\ &\Gamma_3(\mathbf{A}_1), ..., \Gamma_3(\mathbf{A}_6) = \mathbf{A}_1, \mathbf{A}_2, \mathbf{A}_3, \mathbf{A}_4, \mathbf{A}_5, \mathbf{A}_6 \end{aligned} \tag{11.43}$$

where the $\mathbf{A}_K, (K = 1, ..., 6)$, are defined by equation (11.33). Thus the transformation properties of $\mathbf{z}$ are defined by $\Gamma_3(\mathbf{A}_K)$ and we refer to $\mathbf{z}$ as a quantity of type Γ_3.

We first consider the problem of determining functions $\mathbf{z}(\mathbf{x})$ which are linear in $\mathbf{x}$. Upon setting $\mathbf{D}_K = \mathbf{A}_K$ and $\mathbf{R}_K = \mathbf{A}_K$ in equation (11.27), we see that

$$\begin{Vmatrix} \sum_K \bar{A}_{11}^K A_{1j}^K x_j \\ \sum_K \bar{A}_{12}^K A_{1j}^K x_j \end{Vmatrix} = \begin{Vmatrix} 3, & 0 \\ 0, & 3 \end{Vmatrix} \begin{Vmatrix} x_1 \\ x_2 \end{Vmatrix} = 3 \begin{Vmatrix} x_1 \\ x_2 \end{Vmatrix} \tag{11.44}$$

and hence that $\|x_1, x_2\|^T$ is a quantity of type Γ_3. Similarly,

$$\begin{Vmatrix} \sum_K \bar{A}_{11}^K A_{2j}^K x_j \\ \sum_K \bar{A}_{12}^K A_{2j}^K x_j \end{Vmatrix} = \begin{Vmatrix} 0 \\ 0 \end{Vmatrix}, \quad \begin{Vmatrix} \sum_K \bar{A}_{21}^K A_{1j}^K x_j \\ \sum_K \bar{A}_{22}^K A_{1j}^K x_j \end{Vmatrix} = \begin{Vmatrix} 0 \\ 0 \end{Vmatrix},$$

$$\tag{11.45}$$

$$\begin{Vmatrix} \sum_K \bar{A}_{21}^K A_{2j}^K x_j \\ \sum_K \bar{A}_{22}^K A_{2j}^K x_j \end{Vmatrix} = 3 \begin{Vmatrix} x_1 \\ x_2 \end{Vmatrix}$$

yield quantities of type Γ_3. We note upon setting $\mathbf{D}_K = \mathbf{R}_K = \mathbf{A}_K$ in equation (11.28) that the number of linearly independent quantities of type Γ_3 which are linear in $\mathbf{x}$ is given by

$$n_1^* = \frac{1}{6}(2 \cdot 2 + (-1) \cdot (-1) + (-1) \cdot (-1) + 0 \cdot 0 + 0 \cdot 0 + 0 \cdot 0) = 1. \quad (11.46)$$

Given the information furnished by equation (11.46), we see that the computations of equation (11.45) are not required.

The quantities of type Γ_3 which are of degree 2 in $\mathbf{x}$ are obtained upon application of equation (11.30) where we set $\mathbf{D}_K = \mathbf{R}_K = \mathbf{A}_K$. The matrices $\mathbf{A}_K^{(2)}$, i.e. the symmetrized Kronecker squares of the $\mathbf{A}_K$ are given by equation (11.35). The number n_2^* of quantities of type Γ_3 which are of degree 2 in $\mathbf{x}$ is seen from equations (11.29), (11.33) and (11.35) to be given by

$$n_2^* = \frac{1}{6}(2 \cdot 3 + (-1) \cdot 0 + (-1) \cdot 0 + 0 \cdot 1 + 0 \cdot 1 + 0 \cdot 1) = 1. \quad (11.47)$$

We obtain from equation (11.30)

$$\left\| \begin{matrix} \sum_K \bar{A}_{11}^K A_{1j}^{(2)K} x_j^2 \\ \sum_K \bar{A}_{12}^K A_{1j}^{(2)K} x_j^2 \end{matrix} \right\| = \left\| \begin{matrix} 0 \\ 0 \end{matrix} \right\|, \quad \left\| \begin{matrix} \sum_K \bar{A}_{11}^K A_{2j}^{(2)K} x_j^2 \\ \sum_K \bar{A}_{12}^K A_{2j}^{(2)K} x_j^2 \end{matrix} \right\| = \left\| \begin{matrix} 3x_1 x_2 \\ \frac{3}{2}(x_1^2 - x_2^2) \end{matrix} \right\| \quad (11.48)$$

where $\mathbf{x}^2 = \| x_1^2, x_1 x_2, x_2^2 \|^T$. Thus the quantity of type Γ_3 is given by

$$\left\| \begin{matrix} z_1 \\ z_2 \end{matrix} \right\| = \left\| \begin{matrix} 2x_1 x_2 \\ x_1^2 - x_2^2 \end{matrix} \right\|. \quad (11.49)$$

The quantities of type Γ_3 which are of degree 3 in $\mathbf{x}$ are obtained upon application of equation (11.32) where we set $n = 3$ and $\mathbf{D}_K = \mathbf{R}_K = \mathbf{A}_K$. The matrices $\mathbf{A}_K^{(3)}$, i.e. the symmetrized Kronecker cubes of the $\mathbf{A}_K$ are given by equation (11.39). The number n_3^* of quantities of type Γ_3 which are of degree 3 in $\mathbf{x}$ is seen from equations (11.31) and (11.39) to be given by

$$n_3^* = \frac{1}{6}(2 \cdot 4 + (-1) \cdot 1 + (-1) \cdot 1 + 0 \cdot 0 + 0 \cdot 0 + 0 \cdot 0) = 1. \quad (11.50)$$

With equations (11.32) and (11.39), we see that this quantity is given by

$$\left\| \begin{matrix} \sum_K \bar{A}_{11}^K A_{1j}^{(3)K} x_j^3 \\ \sum_K \bar{A}_{12}^K A_{1j}^{(3)K} x_j^3 \end{matrix} \right\| = (x_1^2 + x_2^2) \left\| \begin{matrix} x_1 \\ x_2 \end{matrix} \right\| \quad (11.51)$$

where $\mathbf{x}^3 = \| x_1^3, x_1^2 x_2, x_1 x_2^2, x_2^3 \|^T$.

We observe that we may readily compute the invariants, quantities of type Γ_3, ... provided that we have determined the symmetrized Kronecker products of the matrices $\mathbf{A}_K$. This would be a tedious computation. We are however in the process of writing computer programs which generate the symmetrized Kronecker products and which will carry out all the computations required for the generation of canonical forms for scalar-valued, vector-valued, ... functions of specified degree in the components of a tensor $\mathbf{E}$. We expect to be able to easily generalize the computer programs so as to give canonical forms for functions $W(\mathbf{E}_1, \ \mathbf{E}_2, \ ...)$, $\mathbf{Z}(\mathbf{E}_1, \ \mathbf{E}_2, \ ...)$, ... of several tensors which are invariant under a group $\{A\}$.

References

[1] Xu Y.H., Smith M.M. and Smith G.F., "Computer Aided Generation of Anisotropic Constitutive Expressions", *Int. J. Engng. Sci.* **25**, 711–722 (1987).
[2] Murnaghan F.D., *The Theory of Group Representations*, Johns Hopkins Press (1938).
[3] Lomont J.S., *Applications of Finite Groups*, Academic Press (1959).

A dislocation theory based on volume-to-surface ratio: Fracture behavior of metals

12.1 Introduction

The behavior of a cylinder bar is known to be stiffer than that of a thin plate. In recent times, Sih [1] has shown that the volume to surface ratio, V/A, is fundamental not only in describing material behavior but also in the assessment of material damage. The quantity V/A is indicative of the brittle/ductile behavior of structural components and has been used in the successful design of pipe lines [2] and other applications.

This work is concerned with the equilibrium distribution of dislocations in the plastic zone at the crack tip of a finite specimen [3], the surface of which is traction free. The influence of free surface on dislocations that can be accounted for by the image force tends to enhance plastic flow near the crack front as to affect the fracture behavior. Such an effect was shown to be governed by V/A for different specimen configurations [4]. Volume to surface interaction occurs regardless of whether the solid behaves elastically or plastically. Refer to [5, 6] for the case of small scale yielding. The elastic-plastic fracture behavior of specimens with finite dimensions will be discussed in connection with the role that V/A plays in the process. Dislocation models would be employed to develop relations that explain macrofracture behavior of metals.

12.2 Super-dislocation model

The fracture behavior of metals will be assessed by a model that considers super-dislocation. It involves a dislocation ahead of the crack.

G. C. Sih and E. E. Gdoutos (eds): Mechanics and Physics of Energy Density, 179–193.

Image force. Consider the problem of a center crack of length $2c$ in a finite width plate as shown in Figure 12.1. The equilibrium equations for stresses applied at infinity can be expressed by the force per unit length crack [7]

$$f_c = \frac{1}{2\mu}\left[K^\circ - \frac{\mu B}{\sqrt{2\pi x}}\right]^2 \tag{12.1}$$

and by the force per unit length of dislocation [7]

$$f_d = B\left[\frac{K^\circ}{\sqrt{2\pi x}} - \frac{\mu B}{4\pi}\left(\frac{1}{2x} - \frac{1}{L-x}\right)\right]. \tag{12.2}$$

In equations (12.1) and (12.2), K° is the stress intensity factor in linear elastic fracture mechanics, μ the stress modulus, and L the ligament of specimen from

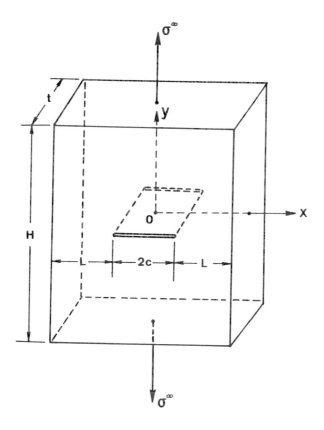

Figure 12.1. Schematic of center crack in plate with finite width.

the crack tip to the edge. Referring to Figure 12.1, the distance x is measured from the dislocation to the crack tip. Both f_c and f_d may also be written as

$$f_c = \frac{k^2}{2\mu}, \qquad f_d = B\sigma_f. \tag{12.3}$$

Here, B, the Burgers vector, applies to a super-dislocation that may contain many elementary crystal dislocations. Denoted by k is an effective local stress intensity factor with effect of dislocation shielding [7] while σ_f stands for the effective friction stress.

Equation (12.1) indicates that the externally applied field is interrupted by a dislocation near the crack tip and contribution of the dislocation to k is $\mu B/\sqrt{2\pi x}$. The term $\mu B^2/8\pi x$ in equation (12.2) can be identified with the image force on the dislocation:

$$f_m = \frac{\mu B^2}{2\pi c} \sqrt{X^2 - 1} \ \left[X + \sqrt{X^2 - 1} \right]^2 - 1 \Big.^{-1},$$

$$\sqrt{2x/c} << 1, \quad \rho < x \leqslant c,$$

$$\tag{12.4}$$

where

$$X = 1 + \frac{x}{c} \tag{12.5}$$

and ρ is the crack tip radius of curvature. The term $\mu B^2/4\pi(L - x)$ in equation (12.2) is the correction for the free surface boundary.

Refer to Figure 12.2 for the stress σ_d associated with f_d on a dislocation near the crack tip. The curve for the total stress is obtained by adding the singular elastic stress; the $1/x$ dependent image stress; and the $1/(L - x)$ image stress correcting for the specimen free surface. Intersections of the yield strength σ_{ys} with the total stress curve give the size of the plastic zone and dislocation free zone as indicated. More specifically, the second and third terms in equation (12.2) cancels out if $L = 3x$. Hence, the main contribution would come from the second and third terms, respectively, for $L > 3x$ and $L < 3x$.

Effective surface energy. Let γ_p be defined as an effective surface energy that includes the predominant plastic work in fracture [5, 6]:

$$2\gamma_p = \frac{1}{\Delta} \int_0^{x_1} f_d \, dx \tag{12.6}$$

in which Δ represents the initial crack growth step that depends on the material microstructure. In what follows, energy dissipated in the form of heat creating irreversibility will be neglected.

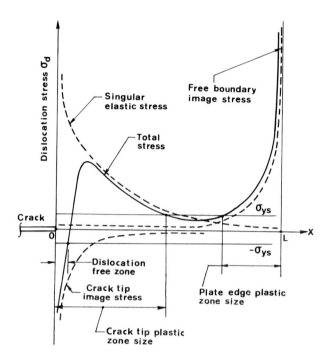

Figure 12.2. Variations of total stress on a local dislocation with distance.

Consider the case where $L \gg x_1$. Retaining only the first term in equation (12.2) yields

$$2\gamma_p = \frac{BK^\circ \sqrt{2x_1}}{\sqrt{\pi}\Delta} \tag{12.7}$$

in which x_1 may be regarded as the size of the crack tip plastic zone. It follows that

$$K^p_{1c} = \sqrt{2\mu G_{1c}} = \left[\frac{2\sqrt{2}\mu BK^\circ}{\sqrt{\pi}\Delta}\right]^{1/2} x_1^{1/4} \tag{12.8}$$

in which $G_{1c} = 2\gamma_p$. The result in equation (12.8) is consistent with that obtained in [5, 6]:

$$K^p_{1c} \sim x_1^{1/4} \tag{12.9}$$

provided that B is the Burgers vector of a super-dislocation. Defining d as the distance where the $x_1^{1/2}$ distribution terminates, K_{1c}^P can be written as

$$K_{1c}^P = \frac{2}{\sqrt{\pi}} K_c^\circ (x_1 \cdot d)^{1/2} .$$ (12.10)

The constant d is related to B only.

12.3 Plastic zone size

The plastic zone size x_p would be affected by the width of the plate and can be estimated by letting $\sigma_f = \sigma_{ys}$ with σ_{ys} being the yield strength. This gives

$$|\sigma_{ys}| = \frac{K^\circ}{\sqrt{2\pi x_p}} - \frac{\mu B}{4\pi}\left(\frac{1}{2x_p} - \frac{1}{L - x_p}\right)$$ (12.11)

which reduces to a quadratic expression in x_p by assuming that $(L - x_p)/x_p = n$, a constant. The possible results are

$$x_p = \frac{1}{8\pi\sigma_{ys}^2}\left\{K^\circ \pm \left[(K^\circ)^2 \pm 2\mu B\sigma_{ys}\left(\frac{1}{n} - \frac{1}{2}\right)\right]^{1/2}\right\}^2 .$$ (12.12)

The positive sign corresponds to the plastic zone boundary and negative sign to that of the dislocation free zone. A number of special cases may be considered.

Infinite medium. For $n \to \infty$, the plastic zone size x_p becomes

$$x_p = \frac{1}{8\pi\sigma_{ys}^2}\{K^\circ + [(K^\circ)^2 - \mu B\sigma_{ys}]^{1/2}\}^2 .$$ (12.13)

Special case of $n = 2$. Setting $n = 2$ in equation (12.12), x_p simplifies to

$$x_p = \frac{1}{2\pi}\left(\frac{K^\circ}{\sigma_{ys}}\right)^2$$ (12.14)

which corresponds to the estimate of yield zone based on the crack tip elastic stress solution.

Plotted in Figure 12.3 are the decay of σ_d the stress corresponding to f_d in equation (12.2) as a function of x for $L = \infty$ and $L = 3x_p$. The plastic zone sizes are the intersections of $\sigma_{ys} =$ const with the curves and the dislocation free

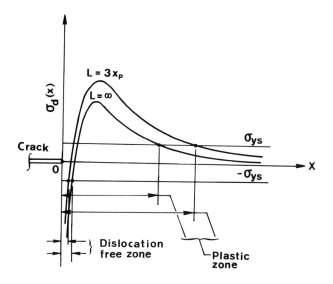

Figure 12.3. Stress on local dislocation influenced by finite plate width.

zone sizes are the intersection of the curves with the line $-\sigma_{ys} = $ const. In general, the plastic zone size tends to increase with decreasing n.

12.4 Dislocation distribution in plastic zone

The distribution of dislocation in the crack tip plastic zone has been studied in connection with the elastic-plastic fracture behavior [10]. The earlier model assumed that the dislocation density function decreased monotonically with distance from the crack tip. That is all dislocations are positive in sign. This is not consistent with the recent experimental results in [11]. Unlike the model in [10], a homogeneous and continuous system of dislocations is considered [12]. Suppose that one of the dislocations with Burgers vector **b** is located at a distance x from a crack of length $2c$. The resultant shear stress on it is zero when the system is in equilibrium. This condition leads to the following integral equation for the dislocation density function $D(x)$:

$$A \int_0^a \frac{D(x')\,dx'}{x - x'} - \sigma_1 + \sigma^\infty + \sigma_0(x) = 0 \tag{12.15}$$

in which σ^∞ is the applied stress at infinity and

$$A = \frac{\mu b}{2\pi(1 - v)}. \tag{12.16}$$

In equation (12.15), σ_1 is the frictional stress of the dislocation motion and $\sigma_0(x)$ is the elastic stress field. The image stresses are left out as they do not alter the qualitative character of the dislocation distribution function. For an elastic-plastic stress field, $\sigma_0(x)$ may be replaced by

$$\sigma_p(x) = \sigma^\infty \left[\frac{1}{\sqrt{1 - X^{-2}}} - 1 \right]. \tag{12.17}$$

where X is given by equation (12.5). The singular integral equation (12.15) can be solved by the finite Hilbert transformation. Applying the condition that $D = 0$ for $x = x_p$, $D(\xi)$ with $\xi = x/c$ becomes

$$A\pi^2 D(\xi) = \sigma_1 \frac{\overline{x_p/c}}{\xi} - 1 \left[2 \frac{\sigma^\infty}{\sigma_1} \tan^{-1} \frac{\overline{2c}}{x_p} + \frac{\pi}{2} + \pi \right]$$

$$- \frac{\sigma^\infty (1 + \xi)}{\sqrt{\xi(2 + \xi)}} \log \frac{\left[\sqrt{2x_p/c} + \overline{(2 + \xi) \frac{x_p}{c} - \xi} \right]^2 + 1}{2 + \frac{x_p}{c} \xi}. \tag{12.18}$$

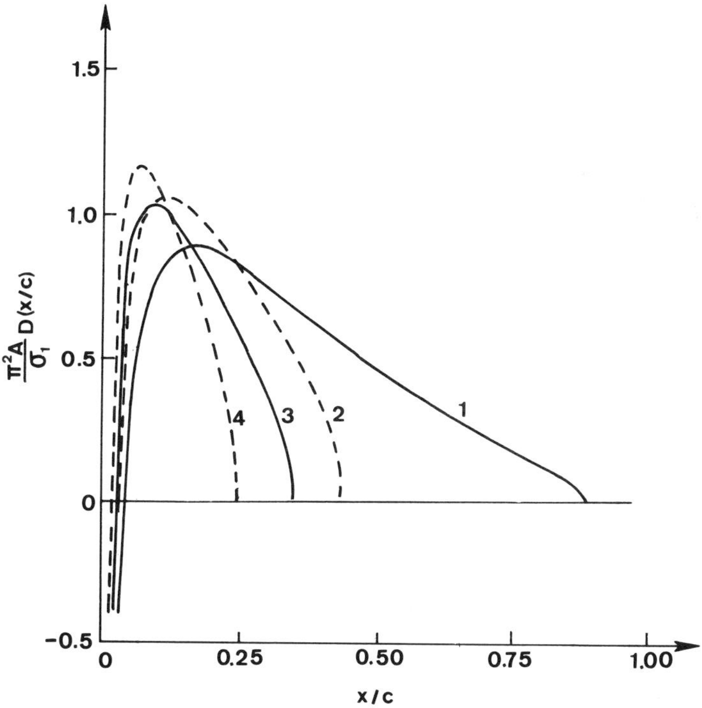

Figure 12.4. Dislocation distribution versus distance for different applied stress and plate ligament width.

The condition for equation (12.18) to exist gives

$$\frac{\sigma^\infty}{\sigma_1} = \pi \left[2\sqrt{\frac{2c}{x_p}} + \cos^{-1}\left(\frac{2c - x_p}{2c + x_p}\right) \right]^{-1}. \tag{12.19}$$

Displayed in Figure 12.4 the variations of the normalized dislocation distribution in the plastic zone with x/c measured from the crack tip. They are shown by the dotted curves 2 and 4 for an infinite medium. The plastic zone size increases with the applied stress $\sigma^\infty/\sigma_1 = 0.48$ (curve 4) and 0.60 (curve 2).

12.5 Crack in semi-infinite medium

To include the image force of dislocation for free surface, the configuration of a semi-infinite medium as shown in Figure 12.5 is considered. Making use of equation (12.15), the equilibrium dislocation distribution in the plastic zone ahead of the crack is governed by

$$A \int_{L-a}^{L} \frac{D_m(x')}{x - x'} dx' - \sigma_p(x) + \sigma^\infty + \sigma_0 + \sigma_m(x) = 0, \tag{12.20}$$

where subscript m is made reference to the dislocation image force. In equation (12.20), $\sigma_m(x)$ is the image force of dislocations acting on the unit dislocation at x given by

$$\sigma_m(x) = A \int_{-L}^{-(L-a)} \frac{D_m(x')}{x - x'} dx' - \frac{A}{2x}. \tag{12.21}$$

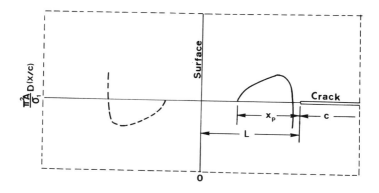

Figure 12.5. Schematic of image dislocation distribution.

An image dislocation distribution may be considered such that the condition of skew-symmetry is satisfied.

$$D_m(-x') = -D_m(x') .$$
(12.22)

Substituting equation (12.21) and (12.17) into (12.20), the resulting singular integral equation can be solved to give

$$\pi^2 A D_m(x) = \sqrt{\frac{x^2 - (L - x_p)^2}{L^2} - x^2} \int_{L-a}^{L} \left[\frac{\dfrac{A}{2x'} + \sigma^\infty + \sigma_p(x) - \sigma_1}{x'^2 - x^2} \right] \times$$

$$\times \left[2x' \sqrt{\frac{L^2 - x'^2}{x'^2 - (L - x_p)^2}} \right] dx' .$$
(12.23)

Equation (12.23) may be solved and expressed in terms of elliptic integrals [3]. Results for yielding influenced by plate boundary surface are shown by the solid curves 1 for $x_p/L = 0.52$ and 3 for $x_p/L = 0.20$ in Figure 12.4 where image force is accounted for. Two applied stress levels σ^∞/σ_1 are considered. They are $\sigma^\infty/\sigma_1 = 0.60$ for curves 1 and 2 and $\sigma^\infty/\sigma_1 = 0.48$ for curves 3 and 4. The former is larger than the latter. Note that a larger difference in the plastic zone size is obtained when the dislocation image force is included. The dislocation image force moves the dislocations in the plastic zone towards the surface and results in larger plastic zones.

12.6 Relation of volume/surface ratio to plate ligament

As the plate ligament L is a parameter that directly influences the dislocation density distribution in the plastic zone, its relationship with V/A can yield insight into why specimens with large V/A tends to behave more brittle while specimens with small V/A acquire more ductile behavior.

For the center crack specimen in Figure 12.1, the V/A ratio is given by

$$\frac{V}{A} = \frac{1}{2} \left[\frac{1}{t} + \frac{1}{2(L + c)} + \frac{1}{H} + \frac{c}{2(L + c)H} \right]^{-1},$$
(12.24)

where t is the thickness. Since $d(V/A)/dL > 0$, V/A decreases as L is decreased. For constant t, c and $G = 2(L + c)$, the relation

$$\frac{V}{A} = \frac{1}{2} \left[\frac{1}{t} + \frac{1}{G} + \frac{1}{H} + \frac{G - 2L}{2GH} \right]^{-1}$$
(12.25)

still gives $d(V/A)/dL > 0$. Hence V/A again decreases with decreasing L Specimens with smaller V/A ratio tend to yield smaller L as it is cracked and results in larger plastic zone. This implies that the specimen would have a more ductile behavior. It is possible to express K_{Ic}^P in terms of V/A by using equation (12.24) or (12.25).

12.7 Specimens with different volume/surface ratio

Specimens with different shapes may have similar behavior if the volume/surface ratios do not differ appreciably. The V/A parameter provides a direct and simple way of determining the brittle/ductile behavior of specimens with different configurations made of the same material.

Circular and square bar. For a circular bar of radius b and length l, it can be shown that

$$\left(\frac{V}{A}\right)^c = l\left[2\left(1 + \frac{b}{l}\right)^{-1}\right]^{-1}.$$
(12.26)

The case of a square bar of length with cross-sectional dimension axa gives

$$\left(\frac{V}{A}\right)^s = l\left\{2\left[1 + 2\left(\frac{a}{l}\right)^{-1}\right]\right\}^{-1}.$$
(12.27)

To achieve the same stress level in both specimens, equate the cross-sectional areas such that

$$b = a + \sqrt{\pi}.$$
(12.28)

In this case for the same length l, Figure 12.6 shows that both $(V/A)^c$ and $(V/A)^s$ varied about the same with a/l. The circular bar tends to have a slightly more brittle behavior than the square bar.

Bundle of fibers. Consider a bundle of n number of fibers with radius b_f and length l. The expression for V/A is given by

$$\left(\frac{V}{A}\right)^b = l\left\{2\left[1 + \sqrt{n}\left(\frac{b}{l}\right)^{-1}\right]\right\}^{-1}$$
(12.29)

where b is the radius of the bundle. Comparing equation (12.29) with equation (12.26) with $\pi n b_f^2 = \pi b^2$, $(V/A)^b$ is much smaller than $(V/A)^c$ as n increases. This

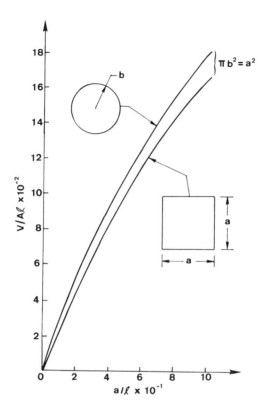

Figure 12.6. Volume/surface ratio as a function of ratio of linear dimensions for solid circular and square bar.

shows that a solid circular cylinder would behave in a much more brittle manner than a bundle of fine fibers of the same length and outside radius. This is because there are considerable more free surfaces in the bundle of fibers.

Circular and square pipe. The V/A ratio for a circular pipe with inner radius b_1 and outer radius b_2 can be written as

$$\left(\frac{V}{A}\right)^p = \frac{1}{2}\left[\frac{1}{l} + \frac{1}{b_2 - b_1}\right]^{-1}.$$ (12.30)

If $\pi(b_2^2 - b_1^2) = b_0^2$ is taken as a constant, then

$$\left(\frac{V}{A}\right)^p = \frac{l}{2}\left\{1 + \left[\sqrt{\frac{1}{\pi}\left(\frac{b_0}{l}\right)^2 + \left(\frac{b_1}{l}\right)^2} - \frac{b_1}{l}\right]^{-1}\right\}^{-1}.$$ (12.31)

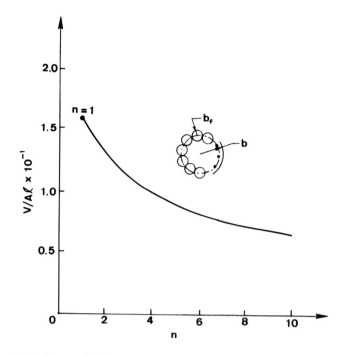

Figure 12.7. Decay of volume/surface ratio with the number of fibers in a bundle.

A plot of $(V/A)^p/l$ versus b_1/l can be found in Figure 12.8. From equations (12.26) and (12.31), it is seen that $(V/A)^p$ is much smaller than $(V/A)^c$ as b_1 increases. A solid circular cylinder behaves more brittle-like than a pipe with a large corner radius.

A square pipe with inner side a_1 and outer side a_2 gives

$$\left(\frac{V}{A}\right)^q = \frac{1}{2}\left(\frac{1}{l} + \frac{2}{a_2 - a_1}\right)^{-1}. \tag{12.32}$$

For constant $(a_2)^2 - (a_1)^2 = (a_0)^2$, equation (12.32) becomes

$$\left(\frac{V}{A}\right)^q = \frac{l}{2}\left\{1 + 2\left[\sqrt{\left(\frac{a_o}{l}\right)^2 + \left(\frac{a_1}{l}\right)^2} - \frac{a_1}{1}\right]^{-1}\right\}^{-1}. \tag{12.33}$$

Variations of $(V/A)^q$ with a_1/l are also displayed in Figure 12.8. The $(V/A)^q$ for the square pipe is much smaller than $(V/A)^s$ in equation (12.27) for the solid square bar in as a_1 increases. The decrease of V/A for a hollow circular pipe with its inner radius is more pronounced than that of a hollow square pipe with its inner opening.

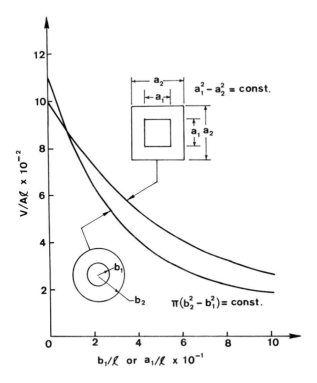

Figure 12.8. Variations of volume/surface ratio with ratio of linear dimensions for hollow circular and square pipe.

Rectangular plate. Plotted in Figure 12.9 are $(V/A)^t$ for a rectangular plate with cross sections $a \times b$ and length l where

$$\left(\frac{V}{A}\right)^t = \frac{l}{2}\left(1 + \frac{l}{a} + \frac{al}{c_0^2}\right)^{-1}.$$

(12.34)

In equation (12.34), $c_0^2 = ab$ is held constant. A maximum of $(V/A)^t$ occurs at $a = b$ which corresponds to the case of a solid square bar.

12.8 Conclusions

A fundamental aspect of this work is the use of dislocation image force to obtain the plastic zones as influenced by the ligament between the crack front and specimen boundary. The volume/surface ratio provides an indication of how the image force would affect the plastic zone size and hence sheds light on

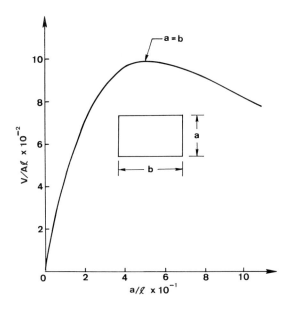

Figure 12.9. Non-monotonic dependency of volume/surface ratio on the ratio of linear dimensions.

the brittle/ductile behavior of specimens. It is desirable to have a relatively low value of V/A so as to enhance ductile behavior. Experiments need to be performed to verify the analytical findings. Positron annihilation would be a good technique [13] for measuring plastic zone size ahead of crack. Load intensity and crack length could be varied for different materials and crack geometries in order to establish the range of validity of the theoretical approach.

Acknowledgements

One of the authors, C.W. Lung, would like to thank Professor Abdus Salam, the International Atomic Energy Agency and UNESCO for hospitality at International Centre for Theoretical Physics, Trieste. This work is supported by the Chinese Academy of Sciences under special grant 87–52.

References:

[1] Sih, G.C., "The role of surface and volume energy in the mechanics of fracture", *Mechanical Properties and Behavior of Solids-Plastics Instabilities,* edited by V. Balakrishnan and C.E. Bottani, World Scientific, Singapore, pp. 396–461 (1985).

[2] Civallero, M., Mirabile, M., and Sih, G.C., "Fracture mechanics in pipeline technology", *Analytical and Experimental Fracture Mechanics,* edited by G.C. Sih and M. Mirabile, Sijhoff and Noordhoff (Now Martinus Nijhoff), The Netherlands, pp. 157–174 (1981).

[3] Liu, S., Xiong, L.Y., and Lung, C.W., "The image force modified dislocation distribution in a cracked finite width material", *Phys. Stat. Sol.* (a) **99**, pp. 97–104 (1987).

[4] Lung, C.W., "The effect of the dislocation image force on the brittle behavior of materials", *J. Phys. D: Appl. Phys.* **20**, pp. 1143–1147 (1987).

[5] Lung, C.W. and Gao, H., *Acta Metallurgica Sinica*, Vol. 19, No. 3, A225 (in Chinese) (1983).

[6] Lung, C.W. and Gao, H., "Analysis of K_{ic} and its temperatures dependence of metals with a simplified dislocation model", *Phys. Stat. Sol.* (a) **87**, pp. 565–569 (1985).

[7] Thomson, R.M., "The fracture of metals", *Physical Metallurgy*, Vol. II, Eds., R.W. Cahn and P. Haasen, Amsterdam: Elsevier, pp. 1487–1551 (1983).

[8] Lung, C.W. and Wang, L., "Image force on the dislocation near a finite length crack tip", *Phil. Mag. A.*, Vol. 50, No. 5, pp. L19–L22, (1984).

[9] Lung, C.W., "Interaction of a dislocation with a crack", *Phys. Stat. Sol.* (8) **85**, pp. K113–116 (1984).

[10] Bilby, B.A., Cottrell, A.H., and Swinden, K.H., "The spread of plastic yield from a notch", *Proc. Roy. Soc.* London, **A272**, pp. 304–314 (1963).

[11] Weertman, J., "Crack tip blunting by dislocation pair creation and separation", *Phil. Mag.* **A43**, pp. 1103–1123 (1981).

[12] Lung. C.W. and Xiong, L.Y., "The dislocation distribution function in the plastic zone at a crack tip", *Phys. Stat. Sol.*, (a) **77**, pp. 81–86 (1983).

[13] Jiang, J., Xiong, L.Y., Lung, C.W., Ji, G.K., Liu, N.Q., and Wang, S.Y., Positron Annihilation (ICPA06), Eds. P.G. Coleman, L.M. Sharma an Diana, (Amsterdam: North-Holland), pp. 469–471, (1982).

The effect of microcracks on energy density

13.1 Introduction

In recent years, there has been considerable interest in the response of polycrystalline brittle solids which contain a homogeneously distributed family of microcracks. As far as the author is aware, the quantification of the effect of such a population of microcracks on the macroscopic response of the solid was initiated by Budiansky and O'Connell [1]. Other contributions have been given in a series of papers by Evans and coworkers [2–6], Hoagland and Embury [7], Clarke [8], Hoenig [9], Laws and Brockenbrough [10, 11] Hutchinson [12] and Charalambides and McMeeking [13].

The topics of major concern in this literature are the loss of stiffness due to a given family of microcracks and the nucleation of further microcracks. In addition, there is considerable interest in the extent to which microcracks in the process zone of a macroscopic crack can have a shielding effect on the macroscopic crack tip. These issues are further discussed in this paper. The development is presented in the same spirit as those of earlier studies [1–13].

All the foregoing contributions refer to *open* microcracks. And, for the most part, the respective authors focus on loss of macroscopic stiffness, rather than change of macroscopic energy density – since the latter (when relevant) is trivially obtained from the former.

At this stage of our discussion it is, perhaps, essential to emphasize a point which is developed by Hutchinson [12] and by Charalambides and McMeeking [13], namely that existence of the standard microscopic energy density does not imply existence of a macroscopic energy density in a microcracking solid. In fact, as will be shown later, severe restrictions must be placed on the microcrack nucleation function in order that the microcracking polycrystalline aggregate can be regarded as macroscopically hyperelastic.

G. C. Sih and E. E. Gdoutos (eds): Mechanics and Physics of Energy Density, 195–201.
© 1992 *Kluwer Academic Publishers. Printed in the Netherlands.*

In addition we draw attention to the important paper of Horii and Nemat-Nasser [14] which pertains to the effects of microcrack closure and friction on the macroscopic response. These authors show that the macroscopic compliance tensor of such a solid depends on the load path and need not even be symmetric.

In this paper, a summary of results is given for loss of stiffness of a solid containing various families of microcracks. For simplicity, we restrict attention to dilute dispersions of microcracks (so that the microcracks do not interact). Further, we make the standard assumption that at least in a first approximation, each grain of the polycrystalline aggregate is isotropic and that residual stresses may be neglected. Next, we discuss some issues which arise from the choice of the microcrack nucleation function. In the absence of definitive experimental evidence, a simple nucleation function is proposed which leads to the microcracked solid being hyperelastic. This choice of nucleation function is not essential but leads easily to an assessment of the extent of microcrack shielding on a macroscopic crack tip. The results thus obtained are compared with some results of Hutchinson [12], Charalambides and McMeeking [13], and Ortiz [16].

13.2 Microcracked solid with given crack density

The approach adopted here closely follows that of Budiansky and O'Connell [1], Laws and Brockenbrough [10], and Hutchinson [12]. Consider an isotropic solid with compliance tensor M_0 which contains a family of variously oriented *open* microcracks. It is essential to assume that the microcracks are homogeneously dispersed. And it is convenient to assume that the microcracks are of similar elliptic planform and that crack orientation and size are not correlated. A convenient measure of microcrack density has been suggested by Budiansky and O'Connell [1]:

$$\beta = \frac{2N}{\pi} \left\{ \frac{A^2}{P} \right\},$$
(13.1)

where N is the number of cracks per unit value, A is the area of a crack and P is perimeter. Also $\{\ \}$ denotes the orientation average of the bracketed quantity.

From [10], we can immediately read off the formula for the loss of stiffness of a solid containing a *dilute* distribution of similar microcracks:

$$M = M_0 + \beta \frac{8E(k)}{3} \{\Lambda\},$$
(13.2)

where Λ is a tensor whose components are given in [10] for a variety of microcrack shapes, and where $E(k)$ is the complete elliptic integral of the second kind.

It is of interest to give some explicit deductions from equation (13.2) for particular microcrack distributions. For a three-dimensional randomly oriented family of penny-shaped cracks, we can recover the Budiansky and O'Connell [1] results

$$\frac{E}{E_0} = 1 - \beta \frac{16(1 - v_0^2)(10 - 3v_0)}{45(2 - v_0)}, \tag{13.3}$$

$$\frac{G}{G_0} = 1 - \beta \frac{32(1 - v_0)(5 - v_0)}{45(2 - v_0)}, \tag{13.4}$$

where G is the shear modulus and v is Poisson's ratio. Also, for a family of aligned penny-shaped cracks

$$\frac{E}{E_0} = 1 - \frac{16}{3}\beta(1 - v_0^2), \tag{13.5}$$

$$\frac{G}{G_0} = 1 - \frac{16}{3}\beta\frac{1 - v_0}{2 - v_0}, \tag{13.6}$$

where E is now Young's modulus normal to the microcrack faces and G is the transverse shear modulus.

When we have a family of aligned slit cracks, we find that the reduction in Young's modulus normal to the slits is given by

$$\frac{E}{E_0} = 1 - \frac{\pi^2}{2}\beta(1 - v_0^2), \tag{13.7}$$

whereas for a two-dimensionally randomly oriented family of slits [10]

$$\frac{E}{E_0} = 1 - \frac{\pi^2}{4}\beta(1 - v_0^2). \tag{13.8}$$

Other explicit results are obtainable but details are not given here.

13.3 Microcrack nucleation

Some potential nucleation functions have been discussed by Brockenbrough and Suresh [15], Hutchinson [12], Charalambides and McMeeking [13] and Ortiz [16]. It is particularly appropriate here to consider the nucleation function in [13] which is intimately associated with an isotropic distribution of three-dimensional randomly oriented penny-shaped cracks. According to Charalambides and McMeeking [13], a good approximation to the Budiansky and O'Connell [1] estimates for stiffness loss is given by

$$\frac{E}{E_0} = \frac{v}{v_0} = 1 - \beta \frac{16}{9}, \tag{13.9}$$

for all crack densities less than 9/16. The tensorial stress-strain relation corresponding to equations of (13.9) is given by Hutchinson [12] in the form

$$\varepsilon = M_0 \sigma - \frac{1}{E_0} \left[1 - \left(1 - \frac{16}{9} \beta \right)^{-1} \right] \sigma. \tag{13.10}$$

Further, the associated nucleation function is proposed for proportional loading to be

$$\beta = f(p), \tag{13.11}$$

where

$$p = (\sigma \cdot \sigma)^{1/2}. \tag{13.12}$$

Thus, under proportional loading, β is taken to increase monotonically until it reaches a saturation value β_s. It then follows from equations (13.10), (13.11) and (13.12) that a macroscopic energy density, U, exists. In fact, for the dilute case

$$U = \frac{1}{2} \sigma \cdot M_0 \sigma + \frac{16}{9} \frac{1}{E_0} F(p) \tag{13.13}$$

where

$$F'(p) = pf(p), \tag{13.14}$$

so that

$$\varepsilon = \frac{\partial U}{\partial \sigma}. \tag{13.15}$$

While it is true that equation (13.9) gives a good approximation to the Budiansky and O'Connell [1] results over a large range of crack densities, it is equally true that the approximations embodied in equation (13.9) are not necessarily good for small values of β – as is evident from equations (13.3) and (13.4). This observation has prompted the investigation of a different nucleation function which closely approximates the form in equation (13.11) but which also leads to the existence of a macroscopic energy density function.

The nucleation function proposed for investigation here is obtained by demanding that the macroscopic stress-strain relation in equation (13.2) be derivable from a macroscopic energy density $U(\sigma)$. Indeed, it is not difficult to show that this requirement implies that the nucleation function under proportional loading is

$$\beta = g(t), \tag{13.16}$$

where

$$t^2 = \sigma \cdot \{\Lambda\} \, \sigma. \tag{13.17}$$

Also the macroscopic energy density is here

$$U = \frac{1}{2} \sigma \cdot M_0 \, \sigma + \frac{8 E(k)}{3} G(t), \tag{13.18}$$

where

$$G'(t) = t g(t). \tag{13.19}$$

Clearly the nucleation function in equation (13.16) is intimately bound up with the stress-strain relation in equation (13.2) and vice-versa. At this juncture, we are content to observe that the proposed function in equation (13.16) exhibits the same anisotropy as the stress-strain relation.

We note that other nucleation functions which do not lead to a macroscopic energy density function have been considered by Hutchinson [12].

A particularly useful consequence of the existence of an energy function is that we can easily determine the extent of shielding on a stationary macroscopic crack due to microcracking in the process zone. This is readily accomplished using the J integral. Following Hutchinson [12], we only consider Mode I small-scale microcracking. Repeating an argument of Evans and Faber [3], Laws and Brockenbrough [11], and Hutchinson [12], we assert that for a contour remote from a macroscopic crack in an isotropic solid

$$J = \frac{(1 - v_0^2) K^2}{E_0}, \tag{13.20}$$

where K is the "applied" stress intensity factor.

On the other hand, for contours in the saturation zone,

$$J_s = \frac{(1 - v_s^2) K_{tip}^2}{E_s}.$$

(13.21)

Invariance of J gives a formula for K_{tip} for isotropic materials.

Thus for a randomly oriented distribution of penny-shaped cracks we find

$$\frac{K_{tip}}{K} = 1 - \frac{8\beta_s}{45(2 - v_0)}(10 - 3v_0 + 8v_0^2 - 3v_0^3).$$

(13.22)

For anisotropic distributions of microcracks, the algebraic details are more tedious. Omitting details, we assert that for a distribution of aligned penny.shaped cracks in the process zone

$$\frac{K_{tip}}{K} = 1 - \frac{16 - 7v_0}{6(2 - v_0)}\beta_s.$$

(13.23)

Also for aligned slit cracks

$$\frac{K_{tip}}{K} = 1 - \frac{3\pi^2}{16}\beta_s,$$

(13.24)

whereas for two-dimensionally randomly oriented slit cracks

$$\frac{K_{tip}}{K} = 1 - \frac{\pi^2}{8}\beta_s.$$

(13.25)

The preceding formula are readily compared and contrasted with the formulae obtained for different nucleation functions by Hutchinson [12], Charalambides and McMeeking [13] and Ortiz [16].

Acknowledgement

The work reported here was supported by the Air Force Office of Scientific Research under Grant Number AFOSR-88-0104.

References

[1] Budiansky, B. and O'Connell, R.J., "Elastic moduli of a cracked solid," *Int. J. Solids Structures,* **12**, (1976), 81.

[2] Evans, A.G., "Microfracture from thermal expansion anisotropy – 1. Single phase systems," *Acta. Metall.,* **26**, (1978), 1845.

[3] Evans, A.G. and Faber, K.T., "Crack-growth resistance of microcracking brittle solids," *J. Am. Ceram. Soc.,* **67**, (1983), 255.

[4] Fu, Y. and Evans, A.G., "Microcrack zone formation in single phase polycrystals," *Acta. Metall.,* **30**, (1982), 1619.

[5] Fu, Y. and Evans, A.G., "Some effects of microcracks on the mechanical properties of brittle solids – 1. Stress strain relations," *Acta. Metall.,* **33**, (1985), 1515.

[6] Fu, Y. and Evans, A.G., "Some effects of microcracks on the mechanical properties of brittle solids – 2. Microcrack toughening," *Acta Metall.,* **33**, (1985), 1525.

[7] Hoagland, R.G. and Embury, J.D., "A treatment of inelastic deformation around a crack tip due to microcracking," *J. Am. Ceram. Soc.,* **63**, (1980), 404.

[8] Clarke, D.R, "A simple calculation of process-zone toughening by microcracking," *Commun. Am. Ceram. Soc.,* (1984), C–15.

[9] Hoenig, A., "Elastic moduli of a non-randomly cracked body," *Int. J. Solids Structures,* **15**, (1979), 137.

[10] Laws, N. and Brockenbrough, J.R., "The effect of microcrack systems on the loss of stiffness of brittle solids," *Int. J. Solids Structures,* **23**, (1987), 1247.

[11] Laws, N. and Brockenbrough, J.R., "Microcracking in polycrystalline solids," *J. Engng. Materials Tech.,* **110**, (1988), 101.

[12] Hutchinson, J.W., "Crack tip shielding by microcracking in brittle solids," *Acta. Metall.,* **35**, (1987), 1605.

[13] Charalambides, P.G. and McMeeking, R.M., "Finite element method simulation of crack propagation in a brittle microcracking solid," *Mech. of Materials,* **6**, (1987), 71.

[14] Horii, M. and Nemat-Nasser, S., "Overall moduli of solids with microcracks: Load-induced anisotropy," *J. Mech. Phys. Solids,* **31**, (1983), 155.

[15] Brockenbrough, J.R. and Suresh, S. "Constitutive behavior of a microcracking brittle solid in cyclic compression" *J. Mech. Phys. Solids,* **35**, (1987), 721.

[16] Ortiz, M., "A continuum theory of microcrack shielding," *J. Appl. Mechanics,* **54**, (1987), 54.

A.A. LIOLIOS

14

Convex energy functions for two-sided solution bounds in unilateral elastomechanics

14.1 Introduction

The versatility of the strain energy density function in mechanics and physics is already well-known. In addition to the role it plays in the development of continuum mechanics theories, the strain energy density function has been used successfully as a failure criterion in fracture mechanics [1–3].

Another remarkable character of the strain energy density function will be discussed in this work; this is concerned with the determination of upper and lower solution bounds in structural mechanics problems. Such bounds are useful to guide the numerical computations. Methods of deriving two-sided solution bounds for linear, elliptic, equality boundary value problems in mathematical physics can be found in [4–7] and [11–13]. The hypercircle method of Synge [4] has been extended to elliptic and hyperbolic inequality problems in elastomechanics [8, 9, 14]. The governing conditions involve equalities as well as inequalities. Variational equality concepts [10, 11] were used.

In what follows, a bound procedure will be developed by application of a generalized strain energy density function in elastostatics where the behavior can be piece-wise linearized. The convexity of this function is a necessary and sufficient condition for constructing a suitable Hilbert space, in which the two-sided solution bounds are obtained. Some admissible elastic solutions are introduced and bounds are calculated by the finite element method using quadratic programming.

14.2 General problem in unilateral elastostatics

Matrix notation will be used in the "natural" finite element formulation

G. C. Sih and E. E. Gdoutos (eds): Mechanics and Physics of Energy Density, 203–209.
© 1992 *Kluwer Academic Publishers. Printed in the Netherlands.*

instead of tensor notations [8, 15]. If G denotes the equilibrium matrix with G^T being its transpose, then the equilibrium conditions are given by

$$G\sigma = p \tag{14.1}$$

in which σ and p are, respectively, the stress and load vector. The compatibility conditions are

$$e = G^T u \tag{14.2}$$

with e being the total strain and u the displacement vector. If the system response can be piece-wise linearized in a way similar to that adopted in plasticity [15], then the constitutive relation can be written as

$$e = \varepsilon + \mu + \theta. \tag{14.3}$$

In equation (14.3), ε and μ are related to σ and v by the relations

$$\varepsilon = E^{-1}\sigma, \qquad \mu = Bv \tag{14.4}$$

with E being the symmetrix and positive stiffness matrix and B the influence matrix of v on μ. The strain vector θ could correspond to that imposed by thermal effects. Denoted as t is the reactive stress vector of unilateral constraints:

$$t = B^T\sigma - Av - r, \quad r \geqslant 0 \tag{14.5}$$

such that

$$t \leqslant 0, \quad v \geqslant 0, \quad v^T t = 0. \tag{14.6}$$

Referring to equation (14.5), r is the resistance vector and A the interaction matrix that is symmetric and is at least semidefinite positive. The above constitutive relations cover most of the piece-wise linearized behavior arising from nonlinear problems in elastomechanics, including those for structures with elements that are unstable and/or fail by fracture. In the latter case, some of the diagonal entries of A may be non-positive [15].

   The problem consists in finding the set (u, v); (σ, t); (e, ε, μ) that satisfy equations (14.1) to (14.6) for a given set of (p, θ). If the quantities v, t and μ are not present, then the problem of elastostatics governed by equations (14.1) to (14.4) is recovered and the solution involves the determination of σ^E, u^E, e^E and ε^E. If the symmetric and negative definite influence matrix Z associated with μ on σ as in dislocations is introduced, then

$$\sigma = \sigma^E + Z\mu. \tag{14.7}$$

Making use of equations (14.5) to (14.7), there results the linear complementarity problem in v:

$$-t = (A - B^T Z B)v + (r - B^T \sigma^E) \geqslant 0 \qquad (14.8)$$

satisfying the conditions in equations (14.1) to (14.6). Existence of solution is assured [15] when the following condition is satisfied:

$$(A - B^T Z B)\text{: positive semidefinite.} \qquad (14.9)$$

It is assumed that the problem stated by equations (14.1) to (14.6) and, hence, equation (14.8) possesses a solution for which the upper and lower bounds would be established.

14.3 Convexity of strain energy and Hilbert space: elastic system

Let S correspond to the state of an elastic system described by (u, v) or (σ, t) according to equations (14.1) to (14.6) and be considered as an element of a Hilbert space H which can be constructed by a suitable scalar product. As scalar product of the two states S_1 and S_2, define the generalized interaction energy

$$S_1 \cdot S_2 = \varepsilon_1^T E \varepsilon_2 + v_1^T A v_2. \qquad (14.10)$$

Making use of equations (14.1) to (14.6) and the principle of virtual work, alternative formulas for the scalar product in equation (14.10) are obtained. They are

$$S_1 \cdot S_2 = \sigma_1^T E^{-1} \sigma_2 + v_1^T A v_2 \qquad (14.11)$$

and

$$S_1 \cdot S_2 = p_1^T u_2 - \sigma_1^T \theta_2 - (t_1 + r)^T v_2. \qquad (14.12)$$

The symmetric and bilinear form of equation (14.10) is positive definite because of E and A is at least positive semidefinite. As mentioned earlier, A may be nonpositive if the system contains elements that are unstable or fail by fracture. Even in this case, equation (14.10) can be shown to be positive definite when equation (14.9) holds.

The squared norm $\|S\|$ in H is

$$S^2 = \|S\|^2 = \varepsilon^T E \varepsilon + v^T A v = 2W \qquad (14.13)$$

and can be interpreted as twice the generalized strain energy W of the system. It follows that the convexity of the function $W = W(\varepsilon, v)$ is fundamental in constructing the Hilbert space H and establishing solution bounds. To this end, two dual classes of admissible states can be obtained from the hypercircle method [4, 8]:

- *Statically Admissible State.* A state S_s with σ_s and $t_s \leqslant 0$ satisfying the static conditions in equations (14.1), (14.4) and (14.5) is said to be statically admissible.
- *Kinematically Admissible State.* A state S_k with u_k and $v_k \geqslant 0$ satisfying the kinematic conditions in equations (14.2) to (14.4) inclusive is said to be kinematically admissible. It can thus be concluded that the sets H_s of all S_s and H_k of all S_k are two convex subsets in H having as a common element the solution state S.

14.4 Global solution bounds

Using the properties of the admissible states, it can be proved [8, 9] that for any pair (S_s, S_k) and the solution state S the following bivariational inequality holds:

$$(S - S_s) \cdot (S - S_k) \leqslant a, \qquad a = -v_k^T t_s \geqslant 0. \tag{14.14}$$

This inequality expresses the principle of virtual work in unilateral elastostatics and is equivalent to the inequality

$$|S - C|^2 \leqslant R^2, \tag{14.15}$$

where

$$C = \tfrac{1}{2}(S_s + S_k), \qquad R^2 = a + \tfrac{1}{4} \| S_s - S_k \|^2. \tag{14.16}$$

A geometric interpretation of the inequality in equation (14.15) is that the solution S in the Hilbert space H lies on or inside the hypersphere (C, R). The following two-sided bounds for the generalized strain energy W is thus found:

$$(\| C \| - R)^2 \leqslant S^2 = 2W \leqslant (\| C \| + R)^2 \tag{14.17}$$

and

$$|S^2 - \tfrac{1}{2}(S_s^2 + S_k^2)| \leqslant \tfrac{1}{2} \| S_s + S_k \| \cdot \| S_s - S_k \|. \tag{14.18}$$

Application of the procedure in [9] yields the dual minimum principles of generalized potential energy

$$\Pi_p(S) \leqslant \Pi_p(S_k) \tag{14.19}$$

or

$$\min \left\{ \Pi_p(S) = \tfrac{1}{2} \varepsilon^T E \varepsilon + \tfrac{1}{2} v^T A v - p^T u + r^T v | \atop \text{s.t. } \varepsilon = G^T u - B v - \theta, \quad v \geqslant 0 \right\} \tag{14.20}$$

and of generalized complementary energy

$$\Pi_c(S) \leqslant \Pi_c(S_s) \tag{14.21}$$

or

$$\min \left\{ \Pi_c(S) = \tfrac{1}{2} \sigma^T E^{-1} \sigma + \tfrac{1}{2} v^T A v + \sigma^T \theta | \atop \text{s.t. } G\sigma = p, \; B^T \sigma - A v - r \leqslant 0 \right\}. \tag{14.22}$$

Using equation (14.12) for the solution state S, it follows that

$$\Pi_p(S) + \Pi_c(S) = S^2 - p^T u + r^T v + \sigma^T \theta = 0. \tag{14.23}$$

Making use of equations (14.19) and (14.21), another global bounds in functional terms are found:

$$-\Pi_c(S_s) \leqslant -\Pi_c(S) = \Pi_p(S) \leqslant \Pi_p(S_k). \tag{14.24}$$

14.5 Local solution bounds

In order to obtain local (pointwise) bounds for a static $F_s(x)$ or a kinematic $F_k(x)$ field quantity at the point x of the system, an appropriate Green state $\hat{S}$ is introduced [5, 6, 8, 13]. Excluding the singularity point x, this state satisfies equations (14.1) to (14.6) with $p = 0$ and $\theta = 0$. To avoid unbounded norm $\|\hat{S}\|$, the Green state is conceived as the sum of the singular part $\overset{\infty}{S}$, corresponding to singularity in the infinite continuum, and of the regular part $\tilde{S}$, satisfying the boundary conditions of $\hat{S}$:

$$\hat{S} = \overset{\infty}{S} + \tilde{S}. \tag{14.25}$$

Since $\hat{S}$, as for S, cannot be found analytically, statically and kinematically admissible states are thus introduced as follows:

$$\hat{S}_s = \overset{\infty}{S} + \tilde{S}^s, \qquad \hat{S}_k = \overset{\infty}{S} + \tilde{S}_k. \tag{14.26}$$

Applying the procedure in [8], equation (14.12) leads to

$$\hat{S}^s \cdot S = F_k(x) - \hat{\sigma}_s^T \theta - (\hat{t}^s + r)^T v \tag{14.27}$$

and

$$S \cdot \hat{S}^k = F_s(x) - p^T \hat{u}^k - (t + r)^T \hat{v}^k. \tag{14.28}$$

We subtract equations (14.27) and (14.28) and denote

$$F(x) = F_k(x) - F_s(x) \tag{14.29}$$

and

$$F_a(x) = p^T \hat{u}_k + \theta^T \hat{\sigma}_s. \tag{14.30}$$

The Schwarz inequality in Hilbert space H gives the estimate of the product

$$|S_0 \cdot \{S - \tfrac{1}{2}(S_s + S_k)\}| \leqslant \|S - \tfrac{1}{2}(S_s + S_k)\| \cdot \|S_0\|, \tag{14.31}$$

where

$$S_0 = \hat{S}_s - \hat{S}^k = \tilde{S}_s - \tilde{S}_k. \tag{14.32}$$

It follows that

$$|F(x) - F_a(x) - \frac{1}{2}(S_s + S_k) \cdot (\tilde{S}_s - \tilde{S}_k)| \leqslant \frac{1}{2} \|S_s - S_k\| \cdot \|\tilde{S}_s - \tilde{S}_k\|. \tag{14.33}$$

The admissible states $\tilde{S}_s$ and $\tilde{S}_k$ satisfy the conditions [8]:

$$\hat{t}_s = \overset{\infty}{t} + \tilde{t}_s = -r, \qquad \hat{v}_k = \overset{\infty}{v} + \tilde{v}_k = 0. \tag{14.34}$$

Note that the decomposition of equations (14.25) to (14.26) for the Green state leads to finite values for the norm of the difference state in equations (14.32) which finally is used in equation (14.33) for the local bounds. Further, conditions in equations (14.34) state that the Green admissible states correspond to those for the elastic system.

14.6 Concluding remarks

Equations (14.17), (14.18), (14.24) and (14.33) provide, respectively, the global and local two-sided solution bounds of the general piece-wise linearized problem stated by equations (14.1) to (14.6) in elastostatics. The bounds are in energy terms for the admissible states and their calculation is based on the definitions in equations (14.11) to (14.13) of a scalar product and norm in a suitable Hilbert space. For these conditions to hold, the convexity of a generalized energy function is pertinent.

On the other hand, the numerical construction of the required admissible states can be guided by the minimum principles in equations (14.19) to (14.22). The finite element models for the displacement and stress formulation [16] correspond to those for the kinematically and the statically admissible states, respectively. As the minimum principles involve inequalities for the constraint conditions, available computer subroutines for quadratic optimization can be used [10, 15]. By constructing convex combinations of admissible states [11] and minimizing the bounds gaps $\| S_s - S_k \|$ and $\| \tilde{S}_s - \tilde{S}_k \|$, improvements of the bounds can be made. The present procedure can be extended to elastodynamics [14]. To reiterate, the obtainment of bounds is based on the convexity of a generalized strain energy function.

References

[1] Sih, G.C., "Some basic problems in fracture mechanics and new concepts", *Eng. Fract. Mech.,* 5, pp. 365–377 (1973).

[2] Gdoutos, E.E., *Problems of mixed mode crack propagation*, Martinus Nijhoff, Leiden (1984).

[3] Sih, G.C., "Mechanics and physics of energy density theory", *Theor. Appl. Fracture Mechanics,* 4, pp. 157–183 (1985).

[4] Synge, J.L., *The hypercircle in mathematical physics*, Cambridge University Press (1957).

[5] Stumpf, H., "Die Berechung statischer und geometrischer Feldgrößen elastischer Randwertprobleme durch punktweise Eingrenzung", *Acta Mechanica,* 12, pp. 223–243 (1971).

[6] Rieder, G., "Eingrenzungen in der Elastizitäts- und Potentialtheorie", *ZAMM,* 52, pp. T340–T347 (1972).

[7] Laws, N., "The use of energy theorems to obtain upper and lower bounds", *J. Inst. Maths. Applics.* 15, pp. 109–119 (1975).

[8] Liolios, A.A., "Upper and lower bounds in nonlinear and unilateral elastostatics", *ZAMM,* 62, pp. T138–T139 (1982).

[9] Liolios, A.A., "A direct formulation of dual extremum principles in unilateral elastostatics by a generalization of the hypercircle method", *ZAMM,* 65, pp. T348–T350 (1985).

[10] Panagiotopoulos, P.D., *Inequality problems in mechanics and applications. Convex and Nonconvex Energy Functions*, Birkhäuser Verlag, Basel (1985).

[11] Velte, W., *Direkte Methoden der Variationsrechung*, B.G. Teubner Verlag, Stuttgart (1976).

[12] Noble, B. and Sewell, M.J., "On dual extremum principles in applied mathematics", *J. Inst. Maths. Applics.,* 9, pp. 273–299 (1972).

[13] Washizu, K., "Bounds for solutions of boundary value problems in elasticity", *J. Math. Phys.* 32, pp. 117–128 (1953).

[14] Liolios, A.A., "Two-sided solution bounds for the Newmark time-stepping approach in unilateral structural elastodynamics", *Proc. 2nd Nat. Congr. Mechanics*, Athens, June 29–July 1, 1989 (to appear).

[15] Maier, G., "Mathematical programming methods in structural Analysis", in: Brebbia, C. and Tottenham, H. (Eds.), *Proc. Int. Conf., Variational Methods in Engineering*, pp. 8/1–8/32, Southampton Univ. Press (1973).

[16] Fraeijs de Veubeke, B., "Displacement and equilibrium models in the finite element method", in: Zienkiewicz, O.C. and Holister, G.S. (Eds.), *Stress analysis*, pp. 145–197, Wiley, New York (1965).